FROM TWO LINES

The Life Story of Captain Josiah Holman

SILVERBIRD
PUBLISHING

Published in Australia by Silverbird Publishing.

First published in Australia 2025

This edition published 2025

Copyright © Michael P. Webb 2025

Cover design, typesetting: WorkingType (www.workingtype.com.au)

The right of Michael P. Webb to be identified as the
Author of the Work has been asserted in accordance with the
Copyright, Designs and Patents Act 1988.

Unless otherwise stated, none of the content in this book is to be reproduced in any form without the prior written permission of the publisher., nor be otherwise circulated in any form of binding or cover other than that in which it is published and without a similar condition being imposed on the subsequent purchaser.

ISBN: 978-1-7641905-4-1

FROM TWO LINES

The Life Story of Captain Josiah Holman

A GENTLEMAN OF GREAT PRACTICAL EXPERIENCE

MICHAEL P. WEBB

Contents

	Years	Page
Introduction		1
PART 1: The Early Years	**1821-1853**	9
Philippines		11
Brazil		13
Cornwall		22
PART 2: Canada	**1853-1856**	25
Introduction		27
Mounting the Expedition		28
London		30
The *Africa*		37
USA and Montreal		41
The Salmon River		45
The Eaton River		50
The Etchemin River & Leeds		55
The Chaudiere River & Mount Megantic		66
Acton & New York		67
Marmora		76
The 1854 Expedition		74

	Years	Page
PART 3: South Africa, Malacca Straits & New Zealand	1856-1862	83
South Africa		85
Malacca Straits		100
New Zealand – Abbey Farm		103
PART 4: Cadia	1862-1893	123
The Beginning		125
Captain Holman Arrives		141
After Cadia		169
New England		180
Peak Downs Copper Mine		208
Retirement, Pastoralist		234
Last Will and Testament		252
Terminology		254
Acknowledgements		257
Source of Illustrations		258

Josiah Holman

Elizabeth Holman

Introduction

Holman is a famous Cornish family name associated with mining in Cornwall, England. The most well-known are the Holman Brothers of Cambourne who began development and manufacture of mining equipment in the early 1800s and rock drills in the 1880s. But the subject of this book, mining captain Josiah Holman, was from a different family line. Much has been written about the Cornish exodus worldwide and Josiah's travels certainly exemplify that narrative. His extraordinary exploits and those of his family took astonishing levels of endurance and adventurism. Their story gives an insight into this emigration era and the incredible distances some journeyed. Eventually, in the early 1860s, after almost two decades on the move, Josiah and family settled in Australia, in the midst of a huge mining boom. Gold had been discovered just a decade earlier. In his subsequent career, Josiah's impact on mining can be traced from Burra Burra in South Australia to Peak Downs in Central Queensland. There were few mining captains held in such high regard in those turbulent times. He was a Captain of Industry:

> The directors have secured the services of a first-class mining captain – Mr. Josiah Holman — acknowledged to be one of the most experienced men of his class in Australia.
> *Australian Town and Country Journal,* 27 Feb 1875

But what of the man himself?

Josiah's father, James Holman, was born on 29 July 1777, christened on 29 August 1780 in Gwennap and died on 20 July 1866, age 89, quite an incredible achievement during those times. He was described as a Mine Engineer and Mine Agent — the latter best described as a mineral prospector. James Holman married Grace Trenwith in 1804 at Mary Tavy, Devon. Grace was born at Redruth in 1787, her parents were William and Elizabeth Trenwith (nee Woolcock). Grace died on 18 November 1867, age 81. James and Grace lived in a number of Cornish villages, Gwennap, Wheal Fortune, Trevarth and Chacewater where they brought up ten children, John (1805–1866), Jerusha Reed (1807–1848), Elizabeth Jackson (1810–1863), Eliza James (b. 1812), James (1814–1873), William (1815–1863), Jecholiah (1817–1840 about), Josiah (1821–1893), Henry (1823-1870) and Emily (1825–1863). It would have been an interesting life for Josiah growing up in a household with so many older siblings. When he was born, the eldest brother, John, would have probably already started work. John's role, before perhaps joining his father, would have been helping his mother Elizabeth with the children, although the 1851 census lists Elizabeth Penrose as a servant in the household. As one child left for work, the next would take over that responsibility of helping out with the children. From Josiah's writing it is evident that a strong sense of family and Christianity was installed in him.

Josiah was born on 27 September 1821 in the small mining village of Gwennap. In the 18th and early 19th centuries Gwennap parish was the richest copper mining district in Cornwall, and was often referred to as the "richest square mile in the Old World". Josiah was baptised in St Wenappa Parish Church, Gwennap on 14 October 1821 and grew up in a place known as Wheal Fortune within the civil parishes of Breage and Sithney in South West Cornwall. He attended school to age of 14 then followed the family's career path, starting work as a miner. In the 1820s and 30s, education for children in Cornwall was generally not widely available. There were few schools in rural areas and many children had to work from a young age. However, there were some efforts to improve education in the county during this period.

The Government's Factory Act of 1833 required business owners to provide some education for children who worked in their factories, and this led to the establishment of some new schools in Cornwall. There were also schools run by churches and other organisations such as the British and Foreign School Society and the National Society for Promoting Religious Education. It is clear education was considered a high priority within the Holman family.

James and Grace Holman, Josiah Holman's parents, circa 1860s.

It is most likely Josiah joined his father around 1835, working at the Creegbrawse Comsols mines (Comsols mines were a collaboration of mines) near Chacewater, Cornwall. In the 1841 census, he is listed as a copper miner not living at home. He was possibly in Wales gaining further education, experience and training hard to became a mine engineer like his father. On the 4 July 1842 at the Gwennap village small Parish Church, Josiah married Elizabeth Simmons, of Truro, Cornwall, born 13 October 1823. He was 20 and she was 18. Both minors (under 21 years of age), they required special dispensation to marry. He is listed as living in Crowan and Elizabeth in Bell Cottage Penryn, near Truro. Her father is listed as a yeoman. Later that year, Elizabeth Simmons Holman was born on 18 November 1842 in Redruth, Cornwall, just a few miles north west of Gwennap. On 10 October 1844, Emily Holman was born but died a year later on 27 October 1845.

The technological advances that catapulted Cornwall's mining industry certainly assisted capitalists, but also brought huge improvements to miners' safety and welfare. In the half century between 1726 and 1775, the average annual tonnage of ore mined increased fourfold. Soon engine houses of carved

St Wenappa Parish Church, Gwennap,
where Josiah was baptised and married, circa 2000.

stone block and towering brick chimneys began dotting the Cornish valleys and fields, easily identifying the mining site nearby. Mineral discoveries, many known for centuries, were suddenly exposed for entrepreneurs to excavate and export. Demand surged, driven by wars, technological advances enabling the Royal Navy to re-hull their boats and massive development of the railway. Speculators, later referred to as adventurers and sometimes called mine agents, were prepared to risk their capital in uncertain ventures and quick to negotiate with landowners.

The gentry made enough from grazing sheep and cattle to be more than comfortable in their privileged place in society, but now had the prospect of greater wealth from what lay underground. These landowners were not about to give up mineral rights of their properties freely. Generally, they insisted on a 21-year lease with the right to foreclose, if in their opinion, the mine was being worked ineffectually. Some took an interest in the mine development, examining construction of shafts to ensure the safety of miners. No property owner wanted hundreds of miners killed on their land, but more importantly wanted to ensure the land itself was not contaminated. They were entitled to inspect the mine at any time to see it was being maintained in good working order regarding the state of shafts, levels, entrance, etc. They could demand construction of new shafts and entrances, installation of equipment and direct development to certain regions of the mine by stipulating that all effort be applied to the exploration of established reserves. In return for permission to work the site the adventurers paid the landlords a proportion of the ore raised, or its monetary equivalent. Others, with little interest, were happy to enter into a royalty arrangement and live off the land as they had always done. Adventurers, those with an entrepreneurial spirit, and the risk takers became more motivated as copper prices, languishing for many years around £70 per ton, began their steady climb to over £200.

But all this was not cheap. Huge sums of capital needed to be raised in hope of great returns which may not be realised for a number of years. Given all this, it is plain that even during mining booms, there would be few winners and many, many losers. The Cornish copper mines soon became deep hard-rock operations, resulting in escalating costs associated with the topography as well as the continuous threat of flooding deep below the surface. Technology initiatives, such as bucket and plunger pumps powered by steam engines soon proved effective in water removal from the deep mines. The considerable cost of these machines dictated the need for large capital investment, thus changing the role of the mine agent. While one eye was kept on the mining operation itself, the other eye watched the fluctuating metal prices.

This was the environment in which Josiah Holman began his apprenticeship years, understudying others and gaining a wealth of experience from some of the great mining captains in Cornwall and overseas. He aimed to follow in his father's footsteps and one day be promoted to a mining captain. The mining captain was responsible for management of the mine itself — often assisted by a purser who prepared the accounts for presentation at quarterly or half yearly meetings of the adventurers or companies. Larger mines often had other managers such as below managers (underground) and above or grass managers. As well as managing the day-to-day operation of the mine, the captain also needed to map out its future, including the cost of such development and likely return on that investment. This required years and years of experience and expertise in the field. Most gained this by rising through the ranks and experiencing different mining operations, seeing what worked and what did not. They needed to be surveyors, with only basic instruments, they needed to be geologists, with no formal education and cost accountants to compute many variables to determine a likely outcome. They needed to be health managers to ensure the safety of the men in their charge, they needed to be HR managers to hire engineers to maintain equipment, boiler men, carpenters for surface and underground work, storemen, labourers and pursers to assist with costing and clerical administration.

Josiah, following his father's footsteps, would have begun his mining career at the very bottom. He would have had some considerable knowledge from his father and elder brothers which would have stood him in good stead. However, it would have been a very arduous first few years and it is most likely he worked in and around the mines of Gwennap. There were certainly plenty of mines to choose from. With his father's contacts, he would have easily found a mining position.

Married to a miner — an ambitious, industrious, adventurous miner — Elizabeth had experiences typical of the time. It is well documented that thousands of miners travelled far and wide to earn money to keep food on the table for their family and it was certainly Josiah's desire to do so. He once wrote, The hope of making money and again meeting my family in better circumstances sooths the sorrow of being parted and bidding good bye."

Overseas salaries were almost double what you could earn at home, incentive for the betterment of the family, but the wives paid the price. Josiah was not at home for the birth of one, possibly two, of his children and one of their children, Emily, died aged one. Elizabeth was not one to stay at home if it was at all possible to travel with her husband. Obviously, adventurism flowed through her veins also. Two of the children were born overseas while on mining expeditions; Emily Louise in Brazil, Charles on a ship off the Cape of Good Hope.

The extraordinary story of Cornish migration is inextricably linked to the rise and subsequent decline of its mining industry. Skilled Cornish miners had been moving about from the 1700s within Cornwall and then to other parts of the British Isles. In the early 1800s the industrial region of Cornwall and West Devon possessed the most advanced mining expertise in Europe and had begun to export its technology, capital and skilled labour. Cornish miners rehabilitated abandoned mines across Latin America in the 1820s and were the first real hard-rock miners in the USA. They worked lead deposits in Wisconsin and Illinois and copper and lead deposits in Norway and Spain in the 1830s, as well as copper fields in South Australia and Michigan in the 1840s. In the main, the Cornish were hard working, thrifty and always looking for opportunities to provide and see their families do well.

Some family dynasties have volumes of information, letters, documents, amassed, tied with string and safely stored in a trunk. Not so the Holmans, understandably, given their movements around the world. To those many family members and those across the ever-shrinking globe — from country New South

Wales, Cornwall, South Africa and Quebec — who have assisted over the past years with this research, I say thank you. I have contacted many local historians and historical societies in areas where Josiah once spent some time and am grateful for the information that they were able to provide. They not only shared information about the notable people he encountered, but also provided me with a glimpse into the lives and experiences of the early settlers during that time.

In the letter of 11 January 1859 written to John Penrose Christoe, Josiah writes:

I have been abroad six times viz. – to the Philippine Islands, the Brazils, twice to the Canadas and the Native Copper Mines of Lake Superior – to South Africa and lastly to Malacca.

I remember when reading these two lines in the original letter and thinking, could this be really true, given he started work age 14 in 1835? Could he really have travelled to all those countries during that period? Then I began to wonder why. Why did he make those journeys and what did he do there?

While the information in this book is in the main historically correct, in many instances I have triangulated dates, events and places in an effort to ensure accuracy. Sections in italics, as above, are taken directly from his letters, journals and numerous reports he wrote to expand the story in his own words. I have also sometimes made assumptions, which may or may not be correct, but have been deduced from information provided and gathered through research as an adjunct to create the narrative. This is a document of historical chronology, pieced together to create a story of incredible adventurism, courage and survival.

PART ONE
THE EARLY YEARS

Philippines

1846–1847

In his January 1859 letter to John Christoe, Josiah writes about his many overseas travels. The places he visited are in chronological order, as they are in this book. The first are the Philippine Islands.

The festive season of 1845 was tinged with sadness for Josiah and Elizabeth. Their second daughter, Emily, just over a year-old, contracted scarlet fever, and although all that could be done was done, the small frail girl failed to respond. During family Christmas gatherings, James — Josiah's father — mentioned a mining expedition was being planned to explore newly discovered gold and copper areas of the Philippine Islands. There had been very little interest from the West since the archipelago came under Spanish rule in the mid-16th century. The Spanish-owned San Remigio Copper Mining company had begun operation in 1842 and coal had been discovered on some of the outlying Islands. Under General Narciso José Anastasio Clavería Zaldúa (1795–1851) a Spanish Army Officer, laws were introduced to control mining in the Philippines in 1846. Appointed Governor General in July 16 1844, he went about implementing laws to: "Promote the mining industry in these Islands by as many means as possible."

He drew up a regulation of mining which provided the opportunity for foreign mining companies to obtain a concession to mine in certain areas of the Philippine archipelago. Concessions were granted through the General Directorate of Mines and overseen by Inspectors or sub- delegates who had the right to inspect or search the mine at any time. There were many regulations introduced to ensure compliance relating to operation and safety.

Governors and mayors of the provinces were not to obstruct development and were to allow the people to be employed by the mining companies, in which case they would pay two Spanish Reales per month tax to fund the towns that housed the local miners. Mining companies were required to compensate for damages or losses caused by any mining operation. There was however an appeals process through the Government Captain General, the Superintendent of The Royal Treasury and the Inspector of Mines. General Clavería was not about to have his Islands plundered by foreigners, as had happened over the many centuries past.

With new regulations now established for legal and ethical mining and the success of the Spanish mining company, interest increased and an exploration team was assembled in England to explore the large archipelago. Josiah was chosen to join this team; he would earn a lot more on such a venture than working the mines of Cornwall. During this expedition, he learned much about the logistics of travel, working with a mining team away from home, foreign protocols, sourcing provisions and keeping them fresh as well as interaction with local communities.

So, when did this all happen? A third child, John Henry, was born on 12 October 1846 in Redruth, Cornwall, so it is conceivable he may have travelled to the Philippines sometime around mid-1846 and returned the following year. In writing to John Christoe in December 1859, Josiah wrote *... 12 years ago I assisted in a large exploration in the Philippine Islands for gold and copper.*

That would have been during 1847.

Brazil

1847–1850

The second country mentioned in his January 1859 letter is Brazil. Hundreds of Cornish miners travelled to Brazil during the 1840s as British companies such as St. John d'El Rey Mining Company, Gongo Soco, began expanding their copper mining operations. The British-owned St. John d'El Rey Mining Company had, some years earlier, abandoned the mine in the province of Minnis Gerais some 30 miles west of the township of Gongo Soco and 300 miles north of the port of Rio de Janeiro. They continued working other mines, Espirito Santo and Raposos, and having experienced some initial success, explored the area further, finally purchasing the Morro Velho gold mine in 1834. The Morro Velho mine is located about 12 miles from Belo Horizonte and lies low between the rolling hills near Nova Lima.

To re-establish operations, the company needed to raise a significant amount of capital. A report tabled at a general meeting of the proprietors held on Thursday 7 May, 1835 indicated:

> There is room for the erection of 120 stamp heads and ample supply of water to keep them in motion. The estate is well provided with wood and water, the supply of all unlimited. Nothing is wanted to increase the production but capital to be employed in the erection of further stamping mills in the opening out of a regular system of underground works, at present work by open cut.

To raise funds to purchase the property and re-establish the mine with new technology, they issued 6,000 shares at £10 each. With the improved infrastructure, production increased and a dividend was paid to shareholders in 1842.

Captain Thomas Treloar, who had worked in the mines of Cocaes, Gongo Soco — having travelled to Brazil in his early 20s — was appointed head mining captain in mid-1846. He succeeded Captain Verran with the promise to increase production to 3500 tons of ore per month. Crushing the ore produced half to one ounce of gold per ton, anything less would barely keep the mine operational, but more, would provide a dividend for shareholders. But to achieve this, he would need experienced competent mining personnel. Employment contracts were generally for three or five years and competition was fierce for the prized positions available. There were more than 1000 people — British, slaves and locals — engaged in the mining and ancillary operations. Borers, the hard rock miners, were paid £10 per month, about double what they would receive in Cornwall, but it was expected they would raise gold worth over £22, thus ensuring the profitability of the mine. Payment was often made directly into bank accounts in England and career progression was swift as often middle management positions were filled by those well-trained Cornish mining apprentices.

The company had learnt from experience that a stable, happy mining force included wives and children and encouraged the Cornish workforce to bring family with them. A Cornish community was well established with shops selling preserved goods augmented with fresh local food sold at weekly markets. The company was also well stocked with animals and poultry. The mining village was now well established with mine stores, charcoal and timber storage areas, a large 30 ft. square office, a new two-story storehouse 120 ft long and 30 ft wide, a blacksmith, carpenter's shop, stables, explosives shed, hospital, school house and various houses for the mining workforce, both black and white. To maintain a regular native workforce, 24 houses for accommodation had been built during the year. There were three churches, one being used, the other in disrepair and another only partially built.

Reverend Charles Wright had been sent out to uphold Christian values and establish a school for the European families.

Life in the tropics was extremely draining. In the rainy summer season, November to March, temperatures rose at times to over 30 degrees Celsius, but felt more like 36 degrees. Winter was mostly dry and cool in the low 20s, but bitter South Atlantic airstreams could drop temperatures to single figures overnight. The approaching summer was characterised by rising temperatures, rain, cooling evening thunderstorms and high humidity. Many of the Europeans complained as there was little respite in the lower valleys.

The March 1851 census reveals Josiah's daughter Emily Louisa was born in Brazil, most likely in Morra Velho. She was born on 15 November 1849 and christened on 25 December 1850 in the St Wenappa Parish Church, Gwennap. We can assume therefore that the family returned home in late 1850 and if the contract was for the customary three years, Josiah would have travelled to Brazil sometime in 1847 after returning from the Philippines. His 10 years' experience, including overseas work, and the fact he was taking his family with him would certainly have worked in his favour. We know he took his wife Elizabeth, as Emily Louisa was born in Brazil, but whether the other children, Elizabeth now aged five, and John Henry aged one, came too is not known. A Cornish settlement of families was well established in Morra Velho at this time, but It was not uncommon for grandparents to bring up children when mining families were overseas. While listed in the 1851 census as a mine agent, in letters Josiah describes himself as a second agent, an assistant or deputy manager in today's terms. So, it seems his years of hard work were paying dividends and this was his first promotion in his extraordinary mining career. Engaged by the company he would have been responsible for directing the miners, checking amount and grade of the grey ore extracted and assisting the mine manager. This may have involved all aspect of processing the ore for delivery to the coast as well as assisting with major projects determined by the mine captain.

Falmouth, on the southern coastline of Cornwall became a major shipping port, especially for the smaller packets in the early 1800s and later larger clippers serving Europe and the Americas. For some time, it was the first port of call for many ships returning from an Atlantic crossing. The typical packet ship was about 200 ft long, with multiple masts and a bluff-bowed hull that, though lacking the speed and grace of the later clipper ships, could plough through the worst North Atlantic seas with reasonable speed and stability. In good weather, a packet ship could cover 200 miles a day. Being smaller, conditions of course were cramped, although there were varying classes of accommodation. They were called packet ships as their first use was delivering mail (packets) around the world, including Australia.

It is most likely Josiah and Elizabeth left from Falmouth by packet ship on a voyage of up to two months to Rio de Janeiro, their passage paid by the St. John d'El Rey Mining Company. The journey to Morro Velho would have been arduous — more than 300 miles over steep mountainous roads and the continual crossing of rivers and streams with pack mules loaded with trunks. Having possibly rested at Juiz de Fora, they trekked through lush prairies and timbered forests, then up the Serra de Mantiqueira, before reaching Nova Lima. After resting for a few days, replenishing fresh provisions, they continued north 20 miles to Morro Velho over many hills, surprisingly some quite barren. The valley, reasonably flat where the village and quarters were situated, spanned almost a mile long, its width half a mile. A stream wound its way below the almost-barren foothills, so entry to the village crossed a rickety bridge desperately in need of repair. Morro Velho was a company town of Cornish miners happily living in well-constructed company houses for only a few shillings a year. There was of course a "pecking order" in securing these cottages defined by size and additional luxuries such as verandas, picket fences and flowering gardens. The quite substantial village comprised around 2000 folk spread out and up the slopes of the Rio das Velhas Valley.

When Josiah arrived in 1847, there were 542 people engaged in the mine department, 33 Europeans and 509 negroes and natives working the shafts, including borers, stope cleaners, kibble fillers, timber men, carpenters, masons, landers, trammers, blacksmiths, pit men and office clerks. He was one of a select few and would learn quickly on the job. There were of course

THE MORRO VELHO GOLD MINE.

Courtesy of Illustrated London News circa 1849.

various mine captains, storekeepers, cashier and physician. A pretty bungalow had been prepared for their arrival and a maid was waiting to help Elizabeth. It was furnished with English fittings, armchairs of oak, a polished dining table, cabinets, sofas, chest of drawers, mirrors, rugs and stylish curtains. For Elizabeth it would have been almost a life of luxury, but for the extreme heat and humidity of summer. As well, it was very familiar for her, with Cornish culture well established; the social structure, religion, education, recreation activities, festivals and music.

In 1848, the mine raised 61,000 tons of ore and in 1849 this had increased to 67,000. This was partially achieved by replacing the narrow-gauge rail line with a wide-gauge system which allowed the borers to be more productive when quarrying the hard grey ore. Still, the use of footways and interconnecting ladders were the main method of conveying machinery and moving people. Also, in 1849 Captain Treloar upgraded and added new machinery to increase stamping capability and water power. But his plans to substantially increase production were hindered by the lack of a competent workforce. Those experienced in mining work had not increased to meet the demand and in fact had decreased by 13. He wrote that work on one of the mines, Gamba to the north, had been suspended due to lack of a "work force", leaving only Bahu and Cachoeira mines operational. In 1848, Captain Treloar became more frustrated by the lack of competent miners. He had the capacity to process ore that could be mined by 310 borers and in 1849 demand had increased to 400 men. Added to that, mortality rates were high. During some years, 10 percent of the workforce — predominantly slaves susceptible to the climatic extremes — died or were too ill to work. Fifty-three deaths were reported amongst the negro population in the first six months of 1849. Of those, five died from accidents in the mine and the rest, including 10 women and children, succumbed to disease — influenza and diarrhoea being the main cause.

There were claims that the negro slaves were being overworked and poorly treated. The contentious slave issue was gaining more traction in British newspapers as various reports filtered through of cruel and harsh treatment. Dr Thomas Walker was appointed in 1850 to carry out a full investigation on the treatment and conditions of the negro slaves. His report provided little evidence of inhumane treatment:

> The condition and treatment of the Imperial Brazilian Mining Company's negroes were in every respect perfectly satisfactory, particularly as regards their clothing, food and health.

It was generally considered they were no worse off than the impoverished of Britain.

Under head captain Treloar, sub-agent Holman was assigned one of the three mines and went about organising the men and implementing new plans set out by Captain Treloar. The 40 feet pumping wheel to increase water power was under construction and 15 new stampers had arrived from Cornwall, all of which would improve output by over 15 percent. It was hoped the new wheel could also be configured to haul the ore using huge buckets (kibbles) from the shafts, thus eliminating the need for whim animals. Maintenance and any repairs were attended to quickly. Josiah's was very much a supervisory role, either morning or night shift, assisting with the smooth running of the mining operation. Responsibilities ranged from checking tools in and out of the store room, determining when to use gunpowder for blasting and how much, providing guidance for the inexperienced miners as well as pit work in the shafts including safety, ventilation, checking the timber head frames and supports etc. Following the same techniques they had learnt back home, one man would locate a crevice and while holding the steel rod, the other miner would hammer at it, inch by inch driving the rod into the stone. Once the drilled holes had been completed to the appropriate depth, the blasting crew would pack the blasting powder mixed with sawdust and charges. Large ore pieces would be broken up and all loaded into baskets or trams to be taken to the smelter. There was also a responsibility to ensure the quotas of ore mined each day were achieved. Day books needed to be updated at commencement and end of shifts and reports were prepared at the end of each week. There was always more than enough to do at a remote mining site …

"The Under Captains are now advocating the necessity of removing a bar of killas, which obstructs the play of the kibble chain, from one of the inclined planes on the middle Cachoeira. This bar of killas has already more than once caused the breakage of the kibble chains and to guard against the mischief which might result from a repetition of such breakages, I presume I must consent to its removal, though it will load us with 200 tons of killas more than we should otherwise have to contend with." (Capt. Treloar report March 1850) on general maintenance.

"The disperse for native labour has greatly increased owing to the activity with which we are endeavouring to push forward the numerous and important works now in hand: viz: new hauling engine, 24 head stamps, hurry inclined plane, railroad from thence to new stamps, (and which has to be cut out from the side of a precipitous hill), watercourse for the new iron pipes, stone pillars, to carry launders for same over the intervening low ground, new stable, small smithy at the Bahu hauling machine, for repairing the chains, addition to spalling-floor and a powerful wall (now just completed) to support all that portion of it which has been added during the last three years, watercourse (through a tunnel) for new mill as the old one independent of its being too small for our present black population will be deprived of its present supply of water as soon as the new iron pipes shall be erected." (Capt. Treloar report May 1850)

STAMPING-MILL.

Courtesy of Illustrated London News

Before leaving, Josiah spent some time assisting with the supervision of the repairs and installation of new beams for the Bahu mine to a level of safety that would allow continuation of mining activity. Captain Treloar called it "the greatest work ever performed at Morro Velho." In his report, he stated:

> "All the hands employed about it will merit the reward promised to them, for the great care, interest, and industry they have uniformly evinced and for the perils they have undergone — a slip of the foot, breaking of a chain, or letting fall any of the materials or tools, would, in all probability, have been attended with fatal consequences, yet not a single casualty has occurred, the men had steady hands, good heads and fearless hearts."

Cornwall

1850–1853

There was little regret in leaving Morro Velho and returning to Chacewater to family and friends. It was a joyous Christmas with celebrations for the christening of Emily Louisa on 25 December 1850, at the small Gwennap Parish church where Josiah and Elizabeth were married. In the March 1851 census, Josiah and Elizabeth are recorded living in Chacewater with their three children listed as scholars and schoolchildren. Listed also is a servant, Elizabeth Peeters aged 16 — an indication of the status of the family at that time.

Now with considerable experience gained from three years in Brazil, Josiah had no difficulty in finding work. He had only just started on one project nearby in Gwennap when he was contacted by a good family friend Humphry Willyams, who with other adventurers owned the Creegbrawse and Penkevil mines between Chacewater and St. Day in the Parish of Kenwyn. Mr Humphry Willyams of Carnanton, Cornwall — a descendant of one of the oldest families in Cornwall — was well known to Josiah as the Member for Truro. He knew first-hand his experience, ability and mining knowledge. Willyams was elected to Parliament in 1849, representing the borough in conjunction with Mr Montague Smith, later Sir Montague Smith. An extensive landowner, miner, banker, smelter and copper merchant, he was well known in London business circles as an industrial financier. His business undertakings were as extensive as they were ambitious, often celebrated for their perceptive vision. He was also a senior partner in the banking firm of Willyams, Willyams and Co, operating as "The Miners' Bank" at Truro and St. Columb.

After visiting the mines and completing an assessment, Josiah presented a thorough report on the condition and future of the mines and was appointed mine captain. The mine was proving to be profitable, but by adding new technological innovations, improvements could certainly be attained. He was keen to follow recently discovered veins which necessitated mining even deeper underground. Already, water entering the shafts was creating great difficulties in the mine operation, often forcing the mine to close for short periods. Josiah proposed to the board the purchase of a steam-driven Cornish beam engine and ancillary equipment, boiler; flywheel, valve gear and 18-ton balance bob to not only improve efficiency, but also provide a solution for the removal of water as the shafts reached deeper depths. He estimated the total cost to be circa £5,500 plus labour installation costs. The board was initially reluctant, but recognised that by extending the mine operation deeper, greater profits could be achieved.

In May 1851, with the board's approval, he advertised for a 20-inch cylinder engine and 18-ton balance bob for the mine. In 1853, at the Penkevil Mine bi-monthly meeting, on 18 February, he presented the accounts and mine operation reports of the previous year and recommended the appointment of Captain John Blight as underground manager (underground agent), salary £5 5s. per month, which was confirmed. Josiah also provided accounts for the Creegbrawse mine bi-monthly meeting, noting that a lode had been recently discovered 2 ft wide producing 3 tons of copper ore, worth £36 per fathom (fm). At a meeting of the Penkevil mine on 3 May he reported they had encountered a new tin lode worth £30 per fathom and he expected over the next few months they would excavate at least four tons of tin.

PART TWO
CANADA
1853–1856

Introduction

Perhaps for a number of reasons, Josiah kept a daily journal of his expedition to Lower Canada (Quebec). There were many intriguing aspects in his journal, which led me to carry out further research. How was he was appointed, what was the motivation of the people who appointed him and why? What was the objective of this exploration? There were also elements of the expedition, well documented, that amazed me. How did he travel around areas that had only been partially explored? The use of local guides would have to helped in this journey — there would have been maps of sorts, but I'm sure they were very basic. And then there's the logistics of getting from place to place and coordinating supplies, accommodation during this arduous expedition.

We are lucky that the Holman family kept a copy of that journal and it was passed from one generation to another. While the information written in the journal is historically correct, in some instances I have made assumptions, which may or may not be correct, but have been deduced from various information provided and gathered through research as an adjunct to create the narrative.

Copy of Captain Josiah's diary courtesy F.Keable

Mounting the Expedition
1853

On April 15, 1853, a notice appeared in the London Gazette:

> British American Land Company's Offices, No. 35½, New Broad Street, London.
>
> NOTICE is hereby given, that a Special meeting of the General Court of Proprietors will be held at the London Tavern, Bishopsgate Street, on Monday the 2nd day of May next, at one o'clock precisely, for the purpose of assenting to or dissenting from a proposal, which will then and there be submitted, for the sale of a certain portion of this Company's lands to a Mining Association, upon terms and conditions which will be then stated by the Directors; and for other business.
>
> By order of the Court of Directors, William C Prince, Secretary

The British American Land Company (BALC), established in 1832, was what we might call today a property developer. The company was formed with the purpose of purchasing and developing land to sell to settlers. By Royal Charter in March 1834 the directors (often referred to as "the Court") secured a Private Act from the British Parliament allowing them to develop and sell the land they had purchased from the Government. Initially this was some 1330 square miles of Crown Land purchased by BALC in 1834 for £120,000 payable over 10 years, in an area of mostly wilderness that was then referred to as Lower Canada. The original directors consisted of well-established wealthy London business people extremely well connected. With additional financing (via investors) in all they acquired or held options over 1700 square miles (over 1million acres) in and around the South Eastern townships.

There was much excitement in 1852 when gold was discovered along the Magog River near Sherbrooke, Lower Canada. Fuelled by rumours and speculation, BALC shares suddenly surged, pushing up the price, which had been languishing at around £50 to over £84. The financial markets and shareholders now asked if gold existed in other areas, and how rich? The Court of Proprietors was compelled to act.

Around the time of the April 1853 notice, Thomas Devas, a director of BALC, had approached a business associate, Humphry Willyams, well known in London business circles. Devas asked Willyams who he would recommend to undertake a mining expedition in Lower Canada. Willyams had no hesitation in recommending his "confidential mine agent," Captain Josiah Holman, whom he had engaged to manage his Penkevil and Creegbrawse mining interests in Cornwall. Devas contacted Josiah and, after a quick introduction and broad outline of the plan, asked if he would be available to lead a mining assessment expedition to Lower Canada. Captain Holman readily accepted knowing that if the expedition was successful, it would likely open up further work opportunities. Terms and conditions were agreed. Josiah would be paid £500 plus all expenses and would lead a team of six to assist with the exploration. He immediately chose John Skews whom he had worked with before as his second-in-command, knowing he could be trusted in what was expected to be quite a demanding and hazardous venture.

London

11 June to 17 June, 1853

On Saturday, June 11, 1853, Josiah left for London after enjoying a few days away from the mines with family. This gave him time to begin his report and assessment of the Wheal Harriet, Wheal Rae, Creegbrawse and Penkivel Mines located just south of Chacewater. Elizabeth was pregnant and unable to accompany him to Canada at this time. For her, it was nice to have had him home — SAFE. The impending trip worried her, despite his reassurances and knowing many of the mining wives in local villages had husbands travelling overseas, making money to improve conditions at home for their families:

> *... the time of taking a temporary farewell of my wife, children parents & friends has arrived being the most unpleasant part of the business, however the hope of making money and again meeting my family in better circumstances sooths the sorrow of being parted and bidding good bye.**

But still, stories abounded of terrifying storms smashing small packet ships. Elizabeth had recently heard the steam ship Josiah would likely be travelling on — the *Africa* — had run aground a few years earlier near Belfast.

Josiah farewelled Elizabeth and their three children: Elizabeth Simmons (aged 10), John Henry (6), and Emily Louisa (3). He assured them he would be back in time for the birth of the child they were expecting in October, a promise intended to calm Elizabeth's anxiety over his long absence. Waving goodbye, he headed up the cobblestone street on a balmy evening,

---

*   text in Italics taken directly from Josiah's transcribed journal.

a small assortment of mining equipment in his trusty leather bag over his shoulder, his clothes in his valise in hand. Boarding the train at nearby Chacewater, he travelled to Truro and after a short walk found an inn for the night. At the coach office early next morning, John Skews joined him and shared the travel arrangements of the other five miners, who would meet them in Liverpool. The coach stopped to change horses at Liskeard, where they had breakfast and then continued on to Plymouth railway station in time to catch the noon train to London.

Arriving in London at Paddington Station just after 10 p.m., they took a carriage to the Portugal Hotel, on Fleet Street. The next morning, Josiah walked to No.1 Queen Street Place to meet Devas, to whom he was to report while in Canada. At the meeting, they were joined by William Henry Tregoning and Humphry Willyams who provided documents and maps, which gave a good sense of the geography and geology of the area he was to explore. Later, they discussed the various reports of mineral finds and the small mining ventures already in operation. After the meeting, Holman and Willyams walked to a large mining equipment warehouse and purchased a theodolite — a surveying instrument with a rotating telescope for measuring horizontal and vertical angles — £24/13/0 in total, a considerable sum. Josiah then went to Lubbock's Bank on Lombard Street and withdrew £210, which he

A London street scene

would use to pay the £150 passage from Liverpool to New York for himself and the six miners.

Later that afternoon, he and Tregoning made their way down Old Broad Street to the Virginia Coffee House at No.4 Newman's Court, St. Michael, Cornhill. Sometime later, they met Mr F Pryor and Mr Field and took a hackney coach to the Piazza Coffeehouse in Covent Garden, which stood next to the magnificent Covent Garden Theatre. The coffeehouse was often dubbed a "temple of luxury":

> *The Coffee Room here is the most luxurious I ever saw – took some wine – Mr Michael Williams joined our party and freely conversed for an hour with Messrs Pryor & Field – at 10 p.m. went to Evans's rooms where we were entertained with some first rate comic songs, singalong and taking our Grogs at 1/- per Glass – a person may sit the evening and take only one, or a dozen Glasses just to meet his taste & pocket – returned to the Portugal at 2 a.m.*

Evans Music-and-Supper Rooms, from Henry George Hibbert,
Fifty years of a Londoner's life, 1916

Fortunately, the next morning Josiah did not have any early pressing meetings, as the previous night's festivities had taken their toll and left him feeling poorly. A cup of tea and biscuit helped and after he found a most suitable place nearby for breakfast, he called on Pryor in his Crown Court office. With several others present, there was much jovial reminiscing about the previous evening as well as the more serious business of his impending adventure. Afterwards, he joined Mr Shakespeare who had kindly invited him to tea which was very much appreciated. On Wednesday, he attended a meeting and presented a report to Willyams and the other owners of the Cornish mines that he had been managing for the past few years. After considering the report, Willyams proposed granting him a 100 guinea payment for past services and 5 guineas per month salary in future. (One guinea equalled 21 shillings, or £1/1s.).

After lunch, he met Thomas Devas and they walked to the BALC offices at 5½ New Bond Street to attend a 1 o'clock meeting of the Directors of the British American Mining Association (BAMA):

These Gentlemen informed me that I must write my Reports of progress in Canada to Mr Willyams under cover to Mr Devas I Queen St. Place that the latter person may peruse the same and forward it to Truro at the Office. I recd a letter (sealed) of introduction & credit to Mr Galt at Montreal.

After the meeting, he joined Willyams and Devas and they walked to Lubbock's Bank in Lombard Street where £30 on account of BALC was draw down to cover upcoming expenses. After completing the transaction, Josiah shook Humphry's hand, thanked him profusely for the introduction and opportunity and promised both gentlemen he would carry out the tasks required with first-class thoroughness.

I had an introduction to the Directors excepting Mr Devas who it appears is the acting person who I presume his being more acquainted with Mining Matters, and the latter Gentleman has the cooperation and advice of Mr Willyams whose knowledge in Mining affairs are valuable, and it was through the recommendation of Mr Willyams that I was selected in this important expedition, for which I am much pleased, and in return will do my best to promote the Company's interest and do honour to Mr Willyams selection of myself.

That evening, Josiah caught a one-horse hansom cab to Euston Square Station for the 9 p.m. train to Liverpool. Through the platform bustle, with valise and carpet bag in hand, he found his second-class carriage in the Liverpool Mail Express. His compartment contained some tipsy Irish soldiers who gradually nodded off, one by one. Arriving at Liverpool at 4 a.m. he took a hansom cab to the nearby Adelphi Hotel rather than walk unsafe streets at that hour. After a few hours' sleep and a hurried breakfast, he walked quickly to the docks to meet the men who had travelled from Penzance by steamboat. They were in high spirits, excited about their new adventure. Checking they had brought all the required equipment, he accompanied them to the hotel where they would be staying the night. He then called on William Tregoning at the office of his brother John Simmons Tregoning and they headed off to lunch. After discussing arrangements for the next day, he left for an early dinner engagement with family friends, the Martins. Husband and wife were in good spirits when he arrived, although Mrs Martin had lost some teeth on the right side of her mouth: *Mrs Martins having lost some of her side teeth makes her look rather more aged than expected.*

After an enjoyable dinner, Josiah thanked his hosts and spent the rest of the evening watching some amusements and a fireworks' display at the Zoological Gardens in West Derby Road. Had he not stayed so long with the Martins he might also have seen the lions and tigers being fed. Rising early, he wrote to Elizabeth about his progress and the meetings in London. He also wrote to his father James and included an interest ticket for £100 on the Miners Bank as well as adding a short Will to the letter. He then visited merchants further down Lime Street and bought some additional clothes for his trip as well as drawing instruments. At the Post Office, he paid for a 30-shilling Post Office order which he sent to his wife. Then he lunched with William Tregoning, who offered to drop off the letter to his father when he returned to Cornwall. They had just finished lunch when joined by the six miners — Skews, Treweek, Whitford, Jennings, James and Marrall — and they then all headed for the Liverpool docks. They stopped at a hotel along the way to buy refreshments for the journey, wine for dinner and brandy afterwards:

Mr W.H. T. accompanied self and men to the wharf and gave each man good advice during their anticipated absence from home especially to be dutiful and take care of the Captain (myself). I may well say that Mr Tregoning is my best friend.

It was only a short walk to the wharf and from the small hill above the wharves they could see the *Africa* in the far distance anchored in the Mersey River. It stood out amongst the many tall-rigged ships, an impressive 300-ft wooden-hulled steamship of 2300 tons capable of holding 1000 tons of coal, 700 tons of cargo and 200 passengers. Equipped with a pair of side-lever engines of around 814 hp as well as two large rigged sails, the *Africa's* twin paddlewheels boasted a large dimension of almost 38 ft driven by four fluid boilers and 20 furnaces consuming some 80 tons of coal each day. Turning into Strand Street, Josiah saw the docks were alive with merchants collecting goods, passengers looking for their tender or ship and drays laden with goods trying to weave their way through the confusion of people scurrying to greet passengers or bid bon-voyage. There were also swindlers attempting to sell or assist the men in any way, but these were quickly dispatched.

Shaking hands vigorously with William, he bid farewell and thanked him for all his assistance with this journey. He really was a very good, supportive friend. The six men passed luggage and equipment to each other from the dock to the tendered satellite steam tug and then started for the *Africa*. Deckhands assisted transferring the luggage and equipment on board as they climbed the gangplank. Once completed, they registered with the purser who directed them to their cabins where they stowed their luggage. Before the anchor was raised, all passengers were called on deck and each one called and tickets checked thoroughly. Most stayed on deck as the ship slowly headed seaward, some waving to well-wishers on shore.

Got on board & luggage stowed in my Cabin at 5 p.m. we passed the Signal place & fired two Guns announcing our departure. Weather was fine and by 6 O'clock we were all making a good dinner on smooth water.

The *Africa*

18 June to 30 June, 1853

The whole accommodation Josiah found excellent for this class of ship:

The sleeping apartments each contain two berths. The saloons for first and second cabin passengers are on deck Aft and Fore the Engine rooms. The After saloon is very long containing two flights of Tables the whole length of the same, with a roomy walk down the centre. The floor is covered with Brussel carpet, and the walk with an extra carpet in shape of stair carpeting but wider. Over the dining table is a rack for holding decanters and glasses which at dinner time is lowered by screws to within 2 feet of the table. After the cloth etc is removed the same is raised by screws to 3 feet above the Tables. Every passenger has to provide his own wines, spirits etc. When a decanter or bottle of wine is ordered a ticket is given for the same and after dinner if any wine is left in the bottle or decanter the same is put on the rack with the party name affixed on a card to it for his use at a subsequent time.

The saloon is well ventilated and lighted by windows about 2½ feet apart the insertions having beautiful Landscapes and other scenery painted, with gilt mouldings, looking very much like real pictures hung in frames. At the entrance of the Saloon either side are two large mirrors in gilt frames. The saloon will accommodate 100 persons comfortably. Our party number about 70 persons including about 10 Ladies. A great number of the passengers are foreigners viz. Americans, Germans, French, Dutch and Spaniards. The second cabin Saloon is very good with the Sleeping apartments and food; in fact, the whole accommodation is excellent for this class. The tonnage of the Africa is

2300 tons started with 1000 tons of Coals and 700 tons of Cargo. The fresh meat including Beef, Mutton, Poultry, Hares etc are preserved in Ice houses on board ship and will keep good the voyage out.

Crossing the Atlantic

Josiah woke to a foggy morning, the grey granite cliffs of the north coast of Ireland could barely be seen as the *Africa* headed further into the open Atlantic Ocean.

At noon took luncheon – by the time got clear of Land, the wind being fresh a little Sea got up and myself and several other passengers were absent from dinner table through seasickness – I was sick twice during the afternoon.

The weather had moderated the following day, and waking late, Josiah spent most of the day resting. The rest certainly helped, and the next day he… *Ate a good breakfast, played shuffles, felt quite well.*

Further into the Atlantic, the ship began lurching through the tumbling ocean, forcing him to take a double dose of Seidlitz powders which fortunately had the desired effect. Catching up with his men later, he found two of them had been very sick. By the next evening, the ocean had calmed and he shared a jug of punch with his team to commemorate Midsummer's Day.

Now a week into their voyage, seagulls appeared and the banks of Newfoundland could be glimpsed through the foggy haze:

> *Fine morning, several passengers came on deck dressed in their Town clothing, apparently expecting to hop on shore after breakfast although we could not hope to land until evening. Foggy but fine weather. Whilst at dinner the Asia passed in sight within a mile saluted and changed colors – before dinner was past a dense fog came on and being near the sand bar at the entrance of New York Harbour the pilot who had been on board some hours, ordered the Anchor to be dropped being unsafe to attempt to pass the bar and likely to run into some of the many vessels lying about us at Anchor – a great disappointment prevailed throughout with the passengers – the evening was wet and most parties played cards, chess & drafts. Yesterday I won 4/- from a Dr. Scot and today 1/- from another person playing at drafts – went to bed at 12 o'clock.*

It was now 30 June, a fine clear morning when the *Africa* heaved anchor at 4 a.m. and drifted slowly towards the docks. On deck, Josiah observed the foreshore on either side of New York Harbour and determined it was the most wonderful he had ever seen: *… scenery entering and on either side of the New York harbour is as beautiful as any I ever saw if not exceeding.*

The harbour entrance was guarded by stone fortifications and areas of woodland sheltered small villages; below, the foreshore was dotted with wooden cabins. The harbour itself was littered with seemingly every type of vessel; sails, steamships and riverboats. Closer to the city, piers and wharves like fingers from a giant hand jutted from the docks. The south end of Manhattan was filled with ships unloading; all this was duplicated on the East River and on the Hudson steamships and river boats waited to go up the river.

William Simpson, Panoramic view of New York City and Brooklyn, circa 1850.

USA and Montreal

1 July to 10 July 1853

After disembarking, Josiah was pleased not to have to pay any duty, although there was a $2 inspection fee for the miners' tools and the theodolite. That evening, he wandered the streets of New York: ... *took dinner, went up Broadway. Saw Barnum's Museum with Theatre and bearded Woman – the latter having given birth to two children and has a beard and whiskers large enough for any man; she looked ugly in women's apparel, in fact it was like a man's head on a woman's body.*

After the show, he walked to the Crystal Palace, which was still under construction:

Went around the exterior of the Crystal Palace which will be a tasty structure when complete. It is about 1/6 the size of the Crystal Palace of London in 1851. The little space of ground about the palace is tastefully laid out in turf and garden plots. The Croton Reservoir a noble structure in the rear of the palace. Saw a pair of oxen said to weigh 9000 lbs or almost 2 tons each. They are beyond comparison. The finest beasts I ever saw and well developed – but should say that they were overestimated in weight. There was a monster pig in adjoining apartment said to weigh 1300 lbs. This I did not see. A great number of trees are growing in the streets which has a pleasant appearance. The bustle with buses, rail cars etc is only second to London, whilst the houses range from 6 to 8 storeys high. The buildings in Broadway are more ornamental and handsome than any street of houses in London but I do not see any public buildings here to compare with those seen in London.

The next morning, he and the six miners caught a cab to Chambers Street Station for the Hudson River train to Montreal. They had not expected the station to be so large and were somewhat overwhelmed by its grandeur and flurry of activity. Once they found the ticketing office, they arranged for their heavy tools and equipment to be taken to the baggage car. The fare was $6, which Josiah thought fair for a trip of 400 miles to Montreal. Unfortunately, they did not know there were two rail companies with lines to Montreal, and after travelling along the right bank of the Hudson River past Troy they realised their baggage had gone on the other line. They alighted at the first station and walked four miles back to Troy. The next train arrived at 6 p.m. and got them to Burlington, Vermont, around 11 p.m., where they spent the night. Their venture had got off to a less-than-ideal start. It was regrettable this error cost them an extra day; they should have reached Montreal by 7 p.m. that evening. Was this a precursor of the logistic challenges they would encounter in this unfamiliar territory?

Having spent the night in Burlington, the men rose early and caught the 6 a.m. train, this time on the correct rail line to Montreal. At the Rouses Point border crossing, their papers were inspected, and their baggage identified and marked. Arriving at Longueuil Station on the east side of the St Lawrence River at noon, Josiah sent the men to collect the rest of the larger equipment which had been sent via a baggage car. Walking to the wharf, he found the ferry to Montreal and caught a cab to Mr Alexander Galt's BALC office 60 St. James Street (Rue Saint-Jacques). In London, he had been advised he would be reporting to Mr Galt, thus was keen to take the earliest opportunity to get a more detailed understanding of the assignment from him. After formalities, he passed on letters from London, including a sealed Letter of Introduction. Galt outlined reports of mineral discoveries and small mining operations. He was convinced further exploration would reveal additional mineral deposits, the extent of which he was relying on Josiah to determine. He told him he had arranged for a Mr Joseph Pennoyer, BALC general agent in Sherbrooke to accompany him and the team to Sherbrooke. He had also arranged for Josiah to meet Mr John Arthur Phillips and Captain Frances Kent, mineral surveyors, later that afternoon.

Josiah was pleased the frank conversation had provided some important details on the likely difficulties and challenges they would face. Very little infrastructure existed and although a few rail links were in place, in many counties' road networks were not fully developed and some, after heavy rain, were almost impassable. Back at Longueuil Station, the men told him the tools and mining equipment had been offloaded on the border at Rouses Point Station some 20 miles away. With no one on board to claim the equipment at the border crossing the previous evening, inspectors had determined it all should stay at the station. So, Josiah bought a return ticket to Rouses Point and found the equipment and tools intact, although a keg containing pestle and mortar was nowhere to be found. He returned to Longueuil with what he had retrieved, crossed on the ferry and sent the men by carriage — with some of the more valuable equipment — to their accommodation at the fashionable St Lawrence Hall Hotel only a few blocks away at 13 Great Saint James Street.

Having dispatched the men, he caught a cab to meet Mr Phillips from London and Captain Kent from Cornwall, mining and metallurgical experts. Both had been sent out the previous month through the advice of John Taylor and Sons, London, highly regarded mining company promoters and mine managers. They had many common interests, having been acquainted with the same London businessmen with mining pursuits. The next morning, Josiah attended the 11 a.m. service at Trinity Church in St Paul Street, east of Place d'Armes. In the evening, he wrote to Elizabeth and also to his father.

After breakfast, he met Charles Joseph Stewart Pennoyer at reception as arranged. He introduced him to the men, then sent them to Longueuil Station to organise loading their equipment onto the train. He then walked to Alexander Galt's BALC office, caught up on any new developments and collected a letter of instruction which was to cover his mining research expedition. Returning to the Longueuil Station, he checked the equipment was securely loaded and left at 4 o'clock for the 100 mile trip to Sherbrooke.

Arrived at Sherbrooke at 7 O'clock – put up at Mr Cheneys Magog Hotel and the Men at Camerons Hotel – every Inn in the country is termed an Hotel.

Sherbrooke Township

Josiah rose early next morning and, with Mr Pennoyer and the men, headed to the junction of the Magog and St Francis Rivers. Having found a suitable area, he instructed the men to set up the tools to begin some shallow dredging and washing of the sandy gravel river edge. He later met Mr John Cummins, a landowner and entrepreneur closely associated with BALC ... *Saw Mr. Cummins at 8 p.m. who is to have two of our men placed under his direction for exploring the Magog River.* They continued over the next few days exploring the Magog and St Francis Rivers, but were disappointed with their panning results. Now, with a better understanding of the terrain, Josiah spent the evening studying maps for their venture along the Salmon River:

> *Purchased cooking utensils, provisions, blankets etc for our tour on the Salmon River the Miners took the bowls and washed some sand near the junction of the St. Francis on the Magog River.*

He then sent John Skews, his second in command, accompanied by Joseph Pennoyer to Bury to oversee the hiring of the guides engaged by Pennoyer. The guides, welcoming the opportunity to earn extra income, comprised local settlers and Abenaki [indigenous] people familiar with the area and rugged terrain.

Salmon River

11 July to 4 August, 1853

Rising very early Josiah collected the fresh meat (mostly pork), vegetables and biscuits he had ordered the day before for the Salmon River expedition. Whitford and Jennings would stay and continue exploration of the Magog area under the supervision of Cummins. Accompanied by the miners Marrall, James and Treweek, they left Sherbrooke at 8 a.m. and at Bury were joined by Skews and Pennoyer as well as the guides that Pennoyer had engaged. They then all continued on to Gould:

> *We arrived at Gould at 7 p.m. after having been driven over 38 miles of the roughest road I ever saw – travelled by a 4 wheeled stage coach – the last 12 miles was unusually rough. Men pitched their tents on the bank of the river and appeared comfortable in their new and mobile apartments.*

Given their instructions, the men, with guides, headed east up the Salmon River towards Megantic Mountain to map out areas where they were panning and take samples of any minerals of note. Josiah with his guide Bailey headed north by horse five miles across mostly open terrain before reaching dense woodlands. They attempted to continue riding but soon found this impossible through the thick forest. Bailey had indicated earlier he knew of a cabin nearby owned by a French-Canadian couple where they could possibly shelter for the night. Welcomed by the couple they ... *lodged in a French Canadians house – the Wife from Irish parents could speak English – the Canadian houses here generally contain only one room on the ground floor with a timber loft over.*

The lower or ground floor does for all the purposes of cooking, eating sitting and sleeping room. Two beds were in the room visible at bedtime three small beds were drawn out from under the large beds – in these three children were placed. Having drunk a good quantity of Tea which I presume was never in China, I went to bed, in the night I had to walk out of the house to make water for want of convenience within – fleas very plentiful at bed.

Leaving at first light, Josiah and Bailey thanked the family for their hospitality, collected the horses and trekked through the swampy dense forest until reaching Lambton around midday. Josiah paid Bailey, thanked him, then swapped his horse for what he hoped would be a more comfortable horse and cart and continued north towards Québec City. Passing through St Georges on the Chaudiere River, he reached River du Loup just after 2 o'clock the following day having overnight found ... *small house with about twenty people sleeping at night – 2 large & 3 small beds, the other parties lay on the floor which was literally covered – I had great difficulty in getting out of bed not to stumble over a woman & child – an abundance of fleas – got up at 4 a.m.*

He had arranged to meet Mr Richard Oatey, a mining captain from Cornwall ... *at 5 called at Mr. Oatey's house, and was informed that he was at the Mine. I proceeded there, and found him conducting the washings with about 30 men and boys. They were at work about 150 yds. above the junction of the Chaudiere on the de Loup river – their work extended about 200 yds in length and were washing on the West side of the river bed probably extending about 2/3 the width – the Stream being low was dammed on the East side. Mr. Oatey was very free, kind and open, told me everything he knew connected with the washings – however the only thing that benefited me by the visit was that I saw in order to get Gold that the shelf has to be broken – saw several Quartz veins in the bed of the river. Mr. Oatey and the Workmen informed me that they never saw any Gold in the Quartz though. The old Gentleman is of opinion that the Gold is produced from these sources – I do not think so myself from the fact of no Gold being found in the veins – if the veins contained pyrites, I should say they were gold bearers.*

The following afternoon they were joined by Mr Thomas Mackie of St. Sylvester, County Lobiniere, who asked if Josiah could meet him at Leeds to provide an opinion of a number of copper lodes discovered in the area. Josiah left around 5 p.m. when the rain had cleared and drove to Mr Baldoc's at St. Francis, where he spent the night. The next day, passing through Tring and then Forthsay (Saint-Évariste-de-Forsyth), he returned the horse and cart at Lambton. Waking early, he left Lambton at 5 a.m. on foot following a rough track and after three hours reached a small dwelling owned by an Irish couple. They agreed to provide breakfast, and from here he continued south reaching Stornoway at 11 a.m … *got through the woods and swamps by 3.30 p.m. took biscuits on the road and halted about half an hour on this stage, finally got to Gould at 5 p.m. walking 26 miles in 12 hours over the roughest path imaginable – very tired.*

His morning plan to leave Gould was dashed by persistent rain, so he spent the day completing a report to Galt, as well as sending a request to Pennoyer for more provisions; 50 lbs of pork, a bushel each of peas and beans and 200 lbs of biscuits. Wishing to travel up the Salmon River, he arranged for a canoe to be built as a boat was not available. With a small team of boatmen, they headed up the meandering Salmon River arriving at the men's camp around noon only to find the miners were still at Megantic Mountain. Dusk was setting in when the miners returned, … *greatly fatigued with the journey – our party informed me that they had found Gold only in one stream and that in small quantity.*

Folding up the tents before dawn, they gathered their equipment and proceeded up the Salmon River for about 11 miles, making camp just below the junction of the Ditton River. Late in the afternoon, Bailey returned with about 20 lb. of trout caught further upstream, a welcome relief to the diet of pork and biscuits. Many of the men were finding the conditions extremely demanding — having been bitten badly and tormented through the night, they got little sleep:

Sunday morning was ushered in by a blast of oaths from one of our men, through his being badly bitten and tormented throughout the night with Mosquitoes and other flies. In fact I have found the men very uncomfortable and dissatisfied with their camp life – however I expect they will soon get accustomed to this kind of work – as for myself I find the fly's very annoying but by tying a handkerchief over my head and ears and sleeping on my back they do not bite me so bad.

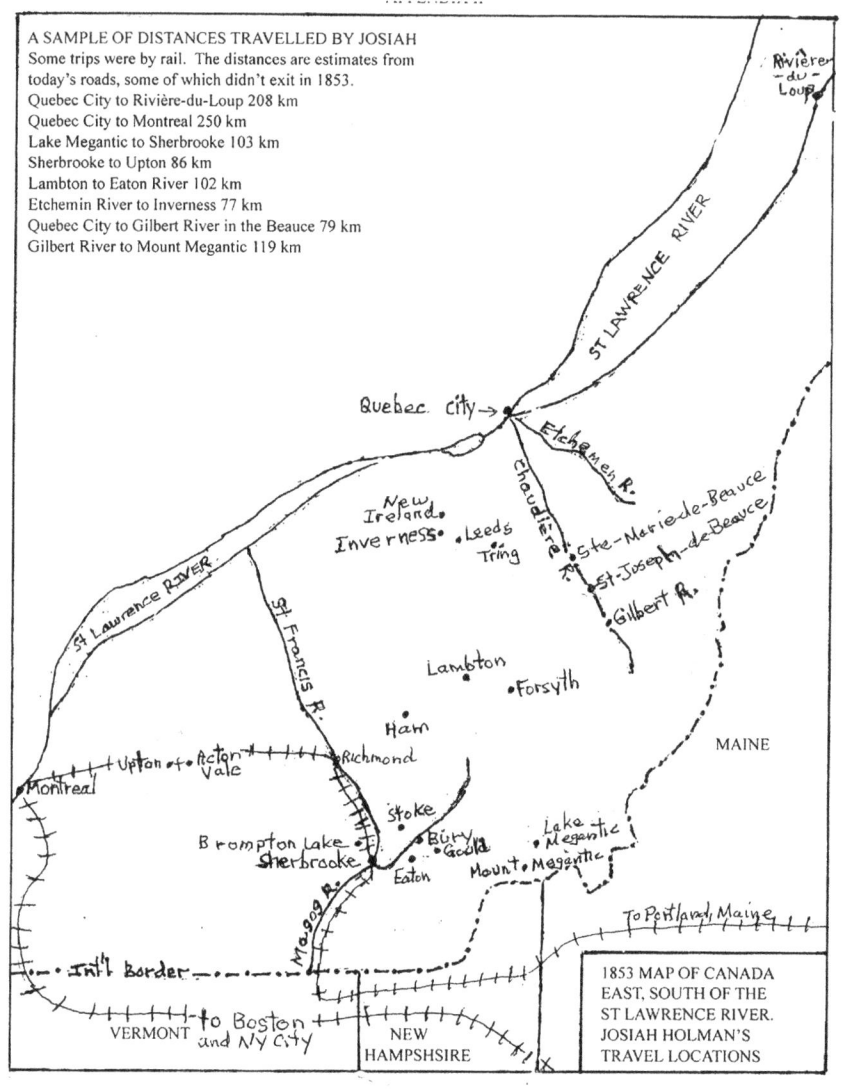

Map provided by Gwen Barry

In the morning, Josiah and a number of the men prepared provisions for a two-day journey, with blankets and necessary light mining equipment:

Went up 7 miles on foot on the river leading to the Megantic Mountain – saw large quantity of Iron ores of a peculiar kind with which I am unacquainted with the proper mineralogical name. Should it be valuable any quantity may be had.

I went up the river 3 miles beyond this – here the bed shows boulders of Elvan, Slate and Quartz but little or no Shelf – at about half a mile above the Diggings there is a considerable fall of water and the shelf is visible, being Slate and I believe Serpentine rock showing numerous small irregular poor Quartz with a little Mundic running at right angles with the Strata, being the first instance that I have noticed in this country.

Breaking camp early the following morning and they continued down the Salmon River arriving at Victoria Falls just before midday. Returning to Gould, Josiah requisitioned a coach for the following morning.

The Eaton River

5 August to 6 September, 1853

Rising early, Josiah organised the schedule for the men; Skews and Treweek would travel with him to Bury and he would then continue on to Sherbrooke with Whitford. The others would make their way to Cookshire and meet on the 8 August to begin exploration of the Eaton River... *Left Gould by Stage at 7 a.m ... arrived at Sherbrooke 38 miles at 4 p.m. accompanied by Mr Gordon & Lady & Mrs. Noble.*

At Sherbrooke Post Office the following morning, he collected £50 from BALC, of which £30 was for his own use to cover expenses and £20 to be paid to Mr Cummins... *Left Whitford here at a Hotel – accompanied Mr Cummins to his house at Roxton falls 6 miles from Acton introduced to Mrs. C and children (6) – Slept here.*

After a hearty breakfast, Cummins took Josiah back to Acton where they met Whitford at the station and all caught a hand rail car, travelling at a very slow 12 miles an hour through heavily wooded countryside to Upton Station. Collecting a horse and cart, they drove some two miles to an area Cummins had identified earlier and after some exploration, having removed the surface covered foliage, some yellow stones could be seen. These produced good specimens similar to the large vein and to Josiah's delight ... *the latter however presents the best features that I have seen in this country.*

Returning to Acton station, he and Whitford paid $1.50 each for a single trip to Sherbrooke. At the Sherbrooke Hotel reception, Josiah collected a letter from Elizabeth dated 19 July, with news of the children and his parents. Rising early, at 7 o'clock they left for Eaton Corner arriving a few hours later and walked a short distance to the Union Hotel where the other men were staying.

The next morning, with horse and cart, they all headed to Colonel Moore's farm, just to the east of the township. Josiah introduced himself to the colonel and was given directions for setting up camp. The men walked a short distance to the Eaton River where they began quarrying large quantities of pyrites. During the following week, Josiah selected places along the riverbed edge where the men could begin digging new pits and washing the sand and alluvial gravel. On Sunday, he would have liked to have found a church to attend but instead wrote a long letter to William Tregoning stating that so far, disappointingly, prospects of finding gold in substantial quantities were not encouraging. He also wrote to Elizabeth and to his father, which he addressed and included in a separate envelope inside the envelope to William.

Wrote Mr Pennoyer stating that 5/6 of our biscuits had gone poor from Mildew although they were left in dry places.

After almost two weeks, Josiah concluded prospects of finding gold along the Eaton River were even less than on the Salmon River. There was gold, sure enough, but from where did it originate?

I believe half mile of the best Ledges seen today would not yield an ounce of Gold, the whole yield of Gold with 9 people today is not enough to weigh in our Scales.

He decided to move on from this disappointing area to an area four miles further up the river, initially along the Clifton branch, then onto Captain Hurd's (Heard) farm. Josiah walked to the main house and introduced himself and was offered dinner and a place to sleep. After dinner, he instructed the men to continue further up the river and wash any sandy beaches they came across and return to the barn for the evening. By then, Bailey had returned from Eaton Corner with a barrel of biscuits.

A man could probably break and wash 15 yards of ledge in a day yielding near 1/2 grains worth, 3d at 4 per ounce after 8 hours hard toil.

Josiah thanked Mrs. Hurd for their hospitality and after breakfast the miners headed towards the boundary line between Lower Canada and the United States. It was hard going through thick forests and resting occasionally, one of the men would climb a tree in the hope of determining an easier route. They found particles of gold in a number of the ledges and continued over the border into New Hampshire for about half a mile, washed some gravel and coarse sand, again not producing any gold nor did they observe any ledges that warranted further exploration.

Josiah had travelled as far east as he wished from Cookshire and had not observed any veins that were likely to contain gold. Even the quartz they had broken up contained only very small specks. They returned to Hurd's farm with the men complaining of heartburn from salty pork and hard biscuits. Skews was the worst, suffering from fatigue and in a poor state. The men continued further examination of the Eaton River while Josiah and Mr Hurd took a rifle and headed into the nearby forest where they shot 13 partridges in about an hour. This was a useful supplement for the men's diet aimed to lift their spirits. Late in the afternoon, a letter arrived from Galt with instructions to examine a copper mine near Québec City; included was a letter of introduction. After a quick breakfast, Josiah instructed the men to head down Eaton River and examine any areas that looked promising, collect samples and make appropriate notes. With Whitford, they headed off at 8 a.m., arriving at Sherbrooke mid-afternoon and checked into the Magog House Hotel. Josiah walked to the BALC office and collected an envelope containing £12/10/- to cover expenses. He then went to Sherbrooke Station where he bought two First Class tickets for tomorrow costing $2.50 each for the trip to Longueuil station.

Waking early, they caught the 6 a.m. train and arrived at Longueuil Station at 10.30 a.m. The steamboat to Québec left at 7 p.m. and reached their destination just after six the next morning, having covered the 180 miles up the St Lawrence River. From the Quebec wharf they caught a cab to the Macromis Hotel, St. Peters Street, and after settling into his room, Josiah took a cab to the Scotch Church for morning prayers. In the evening, he attended the very impressive and stylish Methodist chapel where, once again a collection was required.

With Galt's letter of introduction, Josiah and Whitford walked down the narrow, but impressive, St Peter's Street to No 25, the office of Mr Geo (George) Pemberton, a timber merchant and partner in the firm of Pemberton Brothers. After formal greetings and explanation of his assignment, Josiah was advised that a Mr B. Fisher would accompany him to the copper mine about 18 miles to the east on the Etchemin River. Mr Fisher arrived mid-morning and he, Josiah and Whitford left in a covered Calash carriage to catch the ferry across the St Lawrence River and around 6 p.m. reached the mine site. Lodging had been arranged at a French-Canadian house nearby. In the morning, accompanied by Mr Fisher, they inspected the mine site ... *examined the Mine today showing a small portion of Native Copper – the Vein is not over 3/4 inch wide & not traceable above 3 feet long – it is a worthless spec.*

A farmer having some land adjoining the same reported to Mr Fisher in my presence that he had picked up two pieces of Gold from the surface of his corn field, and produced one of the pieces of Gold – Mr F. commenced washing for Gold selecting samples of earth and shelf, labouring hard for 4 hrs without detecting a particle. I told Mr F. that I expected the man had got it from his son-in-law who had recently returned from California, but Mr. F. could not believe it until he had washed the whole day without finding a spec.

Having no success, they all headed off to the Etchemin River where they broke a number of shelves, again without success. Rising early, they made their way along the Etchemin River to the village of St. Anselme ... *went up the River to St. Anselme village, the bed of the river is wide and shows a great deal of rock, it is variegated clay slate in alternate layers of red, blue and green. At some Mills a little beyond the village, there are some small falls of water and the rock is beautifully exposed, however it does not present anything of interest for mining purposes.*

Mid-afternoon, Josiah left for Quebec by cart, leaving Whitford under Fisher's direction. In the morning, he called on Mr Pemberton at 25 St Peter's Street Québec and reported on the previous two day's activities. Mr Pemberton was of course disappointed to hear his opinion on the likely future operation of the mine ... *in the evening drove to the Marmorenci Falls where the stream falls over a precipice 260 feet deep, the sight is extremely impressive but I regret that one left so late as not to have time to go to the bottom of the falls to witness the change of scenery on the fall. The drive from Quebec, 9 miles to the Falls is very pleasant, passing over a good wood (turnpike) through some neat farms and good grazing land.*

After breakfast, he checked out of the Macromis Hotel and waited for Whitford to return. When Whitford returned, they both headed for the docks and caught the at 5 o'clock *Québec* Steamboat for Montreal. Arriving at Montreal, Josiah caught a cab to Galt's office only to find he was away on business in Upper Canada. As it was not known when he may return, they headed for the wharf, crossed on the ferry and caught the 4 o'clock train arriving at Sherbrooke at 8.30 p.m.

The next day Josiah caught up with Joseph Pennoyer and wife Nancy and all attended morning service at St. Peter's Church on the corner of Commercial and Montreal Streets. An imposing building of brick with impressive oak carved doors, it stood as one of the most prominent buildings in Sherbrooke. Next morning, he walked to the BALC office and after a quick chat with Mr Pennoyer, passed on the completed mine report which was to be forwarded to Galt along with a note he had received earlier from Mr Pemberton. At the Post Office, he collected £250 on account from BAMA and left Sherbrooke at 3 o'clock for Lennoxville, a three-mile journey which, on arrival, gave him time to stroll around the village admiring white washed cottages and fine educational facilities of Bishops College and Grammar School by the river. As arranged previously, he caught up with Chas Pennoyer in the morning. They drove 30 miles by horse and cart to Cookshire where they met the rest of the team who had been making their way downstream exploring the Eaton River. Using part of the £250 he had received the previous day, Josiah paid off Bailey, Mait and the indigenous Abenakis, Annanc and others who had helped them with local knowledge and directions. He then contacted Mr John Baxter, a local carpenter and joiner and organised a two-horse team to take the baggage, mining equipment and the men to Leeds (St Jacques de Leeds) where they were to inspect the area for copper following instructions received through Joseph Pennoyer.

The Etchemin River & Leeds

7 September to 25 September, 1853

After crossing the St Francis River, they found accommodation for the evening: *Slept, paid for Tea, beds and breakfast for 8 persons and feed for horses.* After showers had abated, they spent the day examining some marble quarries with an assortment of colours, changing from almost white to a light speckled grey resembling granite. There was another quarry on a steep hill to the east side some 60 feet from the main road which had been developed to a lesser extent. The quarry of white and grey marble showed signs of blasting: ... t*he party that had the management of the Quarrying department left evident marks in blasting and cleaving that he knew but little in this art.*

In fading light, they examined one more quarry before heading to Mr Rice's lodgings, over very rough roads. Next morning, they headed north for about a mile then turned west off the road for about another mile where they came across a small brook at the foot of Ham Mountain (Mont Ham). Here they broke a number of shelves: ... *not getting any intelligence of veins or Gold being known to exist in this district, we started on the Main Road and before reaching Lake Nicolec saw a pit 4 feet deep by the roadside on Serpentine showing a vein 2 feet wide of Chromate of Iron.* From here they continued just over 3 miles to Wolfstown (Saint-Jacques-le-Majeur-de-Wolfestown) where they found accommodation for the night.

In the morning, they continued towards New Ireland about 10 miles to the north:

The Miners were obliged to walk nearly the whole distance to New Ireland today on account of the rough and hilly state of road.

Here they found accommodation along the roadside almost opposite the Inverness mining property of Dr James Douglas and Co. Dr Douglas and Co. were hoping to raise capital from speculators in London to fund the mining development. Dr Douglas was born in 1800 in Tayside, Scotland and around 1850 acquired properties along the Chaudière River basin of Beauce County. Here he had conducted trials in gold exploration on the Rivière Des Plantes (also known as Rivière Guillaume), and on the Rivière Gilbert, the latter site, Lot 75 of Range 1 North-East was considered to be a rich find. This site was some 4 miles downstream from Richard Oatey's mining operation. Josiah thanked Baxter and paid him 60 shillings for the four days teaming, including taking the baggage and mining equipment eight miles ahead to New Ireland.

The next morning ...*left lodgings taking a guide with us to the Mines of Inverness on lot 4th 2nd range.* Here they examined a shambles of pits, cuttings and mounds of excess and rubble. Josiah first walked to the western pit sunk about 30 ft with a cutting or crosscut driven in from the foot of the hill. South along a vein, about 15 ft below the slope, a shaft had been sunk about 12 to 15 ft below. At the surface, the vein was about 6 ft wide composed of quartz and yielding knobs of yellow grey copper oxide intermixed within cavities of rotten dark green gossan. Walking further, about 12 ft north of the original shaft he discovered a trench about 30 ft long and where it was 3 to 4 ft deep, it opened to the main vein. He had been informed that about 100 lbs of fine copper ore had been mined so far:

Although there are some favourable indications in this Mining property, still on the whole I fear in depth it will not prove well. – this part of the property is in a gentle slope underlying therewith and situated some 30 fms above the brook – but probably full 100 fms from the same rendering crosscutting to tedious.

After completing his notes and having surveyed the mining lease, they walked to Leeds, a five-mile journey staying with William Harrison, a local tinsmith.

Josiah had earlier arranged to meet Mr Lewis Sleeper the following morning. They both set out from Harrison's place in Sleeper's horse and cart to visit copper mines just outside Leeds. Wandering around the mine area, Josiah noted many of the pits were filled with water which restricting examination.

Ordered our Men to drain the pits on the same early the following morning for my inspection. Slept at a house near the Mine – 8 miles from Harrisons.

Meeting with Sleeper at Lot 17 of the 13th Range early next morning, they examined a pit sunk 2 or 3 ft deep and 8 or 9 ft wide displaying a large outcrop of hard crystalline minerals within a quartz vein. *Mr S says he have seen spots of red and yellow copper from this pit – and Gossan is tolerable plenty.* They then moved on to Lot 16 of the 13th Range and examined two small pits which seemed to confirm that the vein previously examined did in fact continue 20 degrees north of east for some considerable length as well. Moving on to the 15th Range, they examined Lot 17 where a number of shallow pits and trenches had been dug. These were more positive, opening up various veins showing stains of copper and fine specimens of rich grey ore. Having spent most the day in the area they left after 4 p.m., arriving at Harrison's just after 7 p.m. After a quick supper, Josiah retired early not feeling well.

Lewis Sleeper would later, 22 December, 1853, purchase the NE half of Lot 17, 15th Range known as Harvey Hill after the Harvey family who had settled in the area around 1830. It became the centre of major copper mining activity in the Leeds area for many years, with more than 15,000 ton of ore reportedly extracted in the next decade or so.

Next morning, Josiah was woken by heavy rain and a storm. Still feeling unwell, he spent the day resting in bed. When the rain cleared, he instructed five of the miners to examine areas either side of Harrison's and with the other miner, set off on horse and cart 15 miles to New Ireland to examine areas that looked suitable for further exploration. Josiah's cold seemed to have worsened and he was also suffering from some conjunctivitis:

Miners out exploring – felt unwell with milk in my eye.

Next day, Sunday, he wrote, *Kept indoors pretty much during the day it is fine weather and for some days past have had grey morning and chilly days giving signs of approaching Winter.*

The following day, he sent four miners to an area beyond Lambie's Mills (later Kinnear's Mills) outside Leeds where he was told copper had been discovered on two of the farms nearby. Finally a bright sunny morning, so Josiah rode to Mr Mackie's mine in Saint Catherine's in the parish of St Sylvester where he caught up with his two miners:

Mr Mackie's proceedings are similar to Dr Douglas and Co. purchasing on speculation thinking to get some London Capitalists to take their dupes through the Agents sent out. I think their designs will be frustrated.

Later he caught up with the four men returning from Lambie's Mills. They had not found any veins containing minerals:

Detained at Harrisons today on account of not being able to get carts to take either my own luggage or Skews – also badly fixed through not having a remittance of cash sent from Sherbrooke by last evenings mail – shall now be obliged to remain here until tomorrow night's post arrives.

Josiah judged the parts of New Ireland, Inverness and Leeds he had seen were ill-adapted for farming purposes, having comprised a succession of hills and narrow valleys of stony, shallow soil. He observed:

Of the many farmers' houses which myself and men have had to put up to, to take dinners whilst out exploring, scarcely a single house would bestow on us a slice of pork or meat of any kind – potatoes, butter and milk with a little fried bread or pancake is the chief food of the majority of poor farmers who call themselves independent. I believe if they would be industrious, that the farms are capable (although the soil is naturally poor) of producing enough to put them in a far better position than they now are. On travelling the country, a labourer is seldom seen in the fields unless about the hay or corn. Fields want clearing of stumps of trees, stones and weeds of many kinds. In tilling time the seed is just muzzled in and not touched till the sickle is wanted and then a good deal is left a fortnight after it is ripe before it is cut.

I am satisfied that about Sherbrooke, Eaton and Limbuck the land is better for farming purposes. A farmer in this country wants capital to begin with for clearing and stocking his farm. The majority of migrant settlers here begin farming without a dollar, consequently it takes them 14 years with industry to get on their legs, in the meantime suffering many privations both in food and clothing leaving out the discomfort of a miserable small dirty log house to dwell in of one room only for all domestic purposes. All poorer class of people let their wives and children go barefooted at home during the Summer and Autumn months.

Walking to Leeds Post Office, he checked if there was a remittance of cash from BALC waiting for him and ...*Finished No 4 Report and copied same into book – posted it as well – wrote to Mr Pennoyer and Pemberton.* The men spent the following day cleaning and packing equipment for next week's venture to the Chaudiere River; Skews and Jennings would be sent to Stoke Mountain to explore and gather mineral specimens. Josiah continued searching various liveries and carters in the hope of being able to secure carts for the trip. Hearing of his requirement, ... *At 3p.m. a carter came here and wanted 6 dollars to take Skews, baggage and tools about 14 miles from this place. After a long debate, he came down to 4 dollars and expenses which I declined giving.*

Exasperated by his unsuccessful search for transport, Josiah approached Harrison who agreed to loan him one of his carts and at 11 o'clock Skews and Jennings left for Stoke Mountain (Monts Stoke) with spare tools, camping equipment and blankets. Josiah instructed the other four men to head off on foot some 30 miles via Broughton to St Joseph's (Saint-Joseph-de-Beauce) on the Chaudiere River and wait for him. As he did each day, he checked at the post office, but still no letter. About to retire for the evening, there was a knock on the door and he was handed a letter which had just arrived, dated 16 September. It had been redirected from Sherbrooke and informed him that Messrs. Galt and Bischoff would be at the Swords Hotel, Québec City on the 27 September and if he was available could he meet them there. The letter also contained a much needed £30.

Once again, he approached Harrison asking if he would be able to secure a horse and wagon for a few days. It took some time to arrange, but finally he and Harrison left at 10 a.m. for St Mary's (St Marie), and then on to accommodation for the evening near St Joseph's — a 25-mile journey. Leaving at 6 a.m., they found Whitford and the other men at their lodgings about four miles upstream from St Francis. Josiah provided instructions to Whitford to begin the next day for Forsyth (Saint-Évariste-de-Forsyth), about 10 miles south and once there, he was to spend the next few days exploring and examining mineral veins in the area.

Chaudiere River & Mount Megantic

26 September to 20 October, 1853

Having completed his assessment of the Leeds area, Josiah was keen to meet with Galt on 27 September. Luckily, he was able to secure a cart to take him 30 miles to Point Levi (Rue Levis) and after crossing the St Lawrence River, checked in at the St George's Hotel St. Peters Street, Lower Town, Quebec City run by Macrow & Son. Walking to the Swords Hotel in the morning he hoped to meet Galt, but was informed Galt was busy in meetings. Josiah returned in the late afternoon and met Charles Bischoff who introduced himself as the BALC Solicitor from London:

> *I saw Mr Bischoff (the BAL Co Solicitor in London) and had some conversation with him respecting our explorations for Minerals in this country – he kindly informed me that he would get a letter from some Gentlemen of Quebec for me to inspect a Mountain said to contain a great quantity of Iron Ores situated near Belville in Upper Canada – this inspection must not interfere with my present engagement – and I am to have a fee for the same.*

Later, Galt joined them and discussed progress and future plans for the exploration. He was in possession of a letter dated 16 August, 1853, from James Bruce, 8th Earl of Elgin and 12th Earl of Kincardine, who was Governor-General of the Province of Canada (1847–54), addressed to Henry Pelham Fiennes Pelham-Clinton, 5th Duke of Newcastle-under-Lyne, Secretary of State for War and the Colonies (1852–1854). Bruce wrote:

Of these Districts, the one which I first visited lies to the south of the St. Lawrence at a distance of about 60 miles from Quebec. The discovery of gold at various points within it, and more particularly in the beds and banks of some of the smaller streams which fall into the river Chaudière, has attracted attention to it of late years. The geological formation in which these discoveries have been made is held to be a prolongation of the Green Mountains of Vermont, and its strata bear a close analogy to those that run through Virginia, the Carolinas, and other Southern States, in which gold has been found at intervals in veins and alluvial deposits.

The gold workings in this District have been hitherto conducted on a very small scale by companies employing hired labour, and for obvious reasons it is difficult, under such circumstances, to ascertain what may have been the amount of the actual yield. The encouragement which the companies have met with is at any rate such as to induce them, after the experience of two or three years, to continue their operations. I was unable to visit the most productive working, but a considerable quantity of gold was extracted in my presence from the gravel on the banks of a small stream called Des Plantes, which runs into the Chaudière.

After some discussion, it was agreed Josiah would explore further the Chaudiere River and Megantic Lake area, then go to Acton — in hope of finding copper veins in the locality. But Galt, having read his previous reports and listened to his opinion of the area generally was concerned they were running out of time to fully explore the area. Josiah wrote, ... *Mr Galt informed me that we should all return to England to Winter and that he had written the Company in London advising them to continue the exploration another Summer season employing myself and only two Miners from England.*

He agreed, knowing his investigation and exploration could be carried out with two men instead of six, at a significantly reduced cost.

Josiah woke early, checked out of the St George's Hotel, posted his letters and returned to the stores he had visited the previous day and bought fresh biscuits and calico cloth for the tents. He also met Mr Otie, his son and son-in-law, who said they had washed 10–11 lbs of gold this season — a similar quantity to the previous year and the year before. Mr Otie believed Dr Douglas's diggings were the most valuable yet discovered in the Chaudiere District. Now running late to meet his carter, Josiah left the men at 10 o'clock and hurried to the ferry to cross the St Lawrence River to Point Levi. The carter he was to have met at 8 o'clock was nowhere to be seen. By late afternoon he decided to engage another carter. They drove about 15 miles, halfway to St Mary's and continued on in the morning, another 12 miles to St Joseph's where he collected the baggage he had left behind the previous Sunday, then a further 10 miles stopping past Tring (Tring-Junction), finally reaching Forsyth (Saint-Évariste-de-Forsyth) at 9 o'clock. He didn't have to search far to find the men and discussed their results while inspecting the pits they had sunk around some exposed veins.

During the recent months of the expedition, having seen so many similar veins which disappeared into nothing, Josiah was quickly coming to the conclusion the area was lacking any mineral value of significance:

I have still an unfavourable opinion of the diggings proving profitable, but think it highly desirable that the Chaudiere & tributaries should have a minute examination for the chance of discovering Gold bearing Veins which no doubt exist on this region.

He was quite sure that many surveys would follow him next summer:

Whilst at Quebec there appeared to be a great stir with the people about the diggings and I expect next season will bring some fresh Companies into the Gold regions.

Josiah and the miners packed the equipment onto the cart and left for Lambton, the men on foot and after travelling some 12 miles made camp just past Lambton at the head of St Francis Lake:

> *Intended going down the St Francis Lake to examine a river leading into the same reported here to show a deal of ledge, however heavy rain prevented us – and I put the men to making our Cotton stuff into shape for a camp for the Megantic determined to wait here till Skews & Co comes up – as the nearest route to the Chaudiere is to strike off from Lampton Village.*

Finally, a sunny morning, they headed a few miles by cart towards Gould, hoping to meet Skews and the men along the way. Returning, he was now concerned Skews might not have received the note he sent to Mr Pennoyer the previous month. Late in the afternoon, snow began to fall heavily, but much to his relief, Treweek with Skews, Jennings and the men turned up, having received his note on 1 October. Returning as instructed, they reached Cookshire on Monday 3 October, and due to heavy rain had not proceeded further. When they were able to leave the following day, ... *they reported the roads to be the worst they have passed over in this country (between Gould & St Francis Lake) – the day was wet when they came here and in the evening it was snowing fast.* Josiah had hoped to proceed to the Chaudiere River but instead was forced to stay at the camp.

Now with only a few inches of snow remaining, they left the camp on a cold and frosty morning and headed east towards the Chaudiere River. The horse and cart carrying their tools and baggage was able to follow a very wet and muddy road made worse by the thawing snow for six miles through mostly settled country. When the road petered out into a single track, the men unloaded, carrying their packs, food and equipment, Josiah his leather bag of hand tools, ramming shovel and gun. After travelling a further three miles, they noticed a small 10-foot square log hut on the edge of a settlement where they stayed the night.

Departing early, Josiah, the six miners as well as five locals engaged to carry provisions and equipment for the three-week expedition, left heavily loaded. While at Lambton, using the money received earlier, Josiah had purchased a barrel of biscuits, 20 pounds of mutton, 13 pounds of sugar, 11 pounds of butter, 9 pounds of pork, tea, salt, pepper and ginger. He would have liked to have purchased more pork, but there was none to be found. It was almost

10 o'clock when, after walking about three miles they came across a small lake (Lac Drolet). By the lake was a canoe, so Josiah with as much baggage as the canoe would take, paddled across, unloaded and returned to collect more equipment while the men walked around the lake. This eased the burden on the men for about 4 hours and they then all continued to the Chaudiere River arriving just after 5 o'clock. The men were exhausted from the day's trekking. Josiah wrote of having ... *carried a Gun & ramming shovel on my shoulders all day which made them sore.* Having pitched the tents by the river's edge, they prepared barley bread and butter for the evening meal.

Next day, Josiah discharged two of the packmen and instructed the others to make a log canoe. He directed the miners to follow a small stream on the west side of the Chaudiere River and to dig and wash any suitable sandy beaches or ledges they may find. Now with better weather, he decided to move on and directed the men to disassemble the campsite and pack all the equipment into the canoe. The men walked along the riverside through wild woods, sharing the load from time to time and at 4 o'clock they found a suitable place to make camp and set about making a good fire to dry themselves. The Canadian guides caught four fish which supplemented the evening meal. It had continued snowing overnight which would make the day's journey as difficult as it was before. They left camp at 9 o'clock and went up river about a mile to rapids that ran steeply through rocky outcrops — requiring the men to push the canoe in the freezing waters. Josiah decided to leave the canoe and asked the men to carry all the equipment on their backs: ... *the day was unusually wet from the falling snow & what also fell from the trees making us all regularly drenched* .

Taking advantage of a fine morning, they left early crossing a small stream almost immediately after leaving camp. The free-flowing rocky river and streams eliminated any possibility of finding deposits of sand or gravel to pan and wash. The journey through the unrelenting wilderness was now becoming increasingly difficult, trekking through dense cedar bush and swamps of decayed timber. At one point, clambering over a fallen tree, Josiah ... *fell down and ran a piece of wood into the heel of my right hand causing a free discharge of blood for a little while – put some leaves to it which had an immediate effect stopping the blood.* Around mid-afternoon they rested and Skews climbed a nearby tree and announced he could see Lake Megantic about six miles in the distance. Josiah calculated they had walked almost 25 miles, through thick unforgiving forest and boggy swamps following the Chaudiere river.

Now late in the afternoon, they came across a narrow road to Victoria, more like a track, but at least not dense bush. Although it was tempting to continue on, Josiah decided to make camp here. The men were keen to continue and rose before light and headed at a quick pace to Gould, arriving at 6 o'clock. They all enjoyed a hearty breakfast having for the past two days … *had no other dinner than biscuits & sugar which we ate by a brook – no spirits of any kind.*

After breakfast, Josiah dismissed, thanked and paid the remaining local guides, and with the miners boarded a Stage Van at 8 o'clock that would take them to Sherbrooke via Bury and Cookshire. It had been slow going travelling the 35 miles over soft appalling, slippery muddy roads.

In the morning Josiah walked to the BALC office hoping to meet up with Alexander Galt. He was informed he was away, but would be returning in a few days. There were letters waiting for him, one from John Tregoning, another from his father and one from his daughter Elizabeth. Trekking for the past week had exhausted him, but the letter from his daughter — her first attempt to write a letter — lifted his spirits immensely. Especially as she had remembered, or was reminded, that it was his birthday on the 27th of last month. There was also a letter asking him to inspect the Marmora Iron works near Bellville on Lake Ontario. He was to undertake this research before he left for Liverpool and would be paid £50, but this would not include any of his expenses.

Acton & New York

20 October to 6 November, 1853

Rising early to see the men before they departed to Dudswell, Josiah then walked to Sherbrooke station and caught the train to Acton (Acton Vale) via Richmond. He had hoped to catch up with John Cummins but was informed he had left earlier for Montréal. Returning quickly to the station, he caught a train to Montréal where, after some searching, he was able to locate Cummins. Returning to Upton together they discussed possible opportunities regarding Cummins copper find.

It was a frosty sunny morning when Josiah and Cummins left, driving by cart 2 ½ miles north of Upton to view the copper veins. Investigating further, they were impressed with the specimens of rich yellow ore stained with green carbonate of copper near the limestone surface. Josiah wrote, ... *it is probable that 2 or 3 weeks labour will either make a great change in the speculation either for better or worse.* The proprietor of the land they were exploring was from Cornwall and Josiah was confident that he would allow such exploration to take place on his property. Nevertheless, he remained pessimistic about the ore: ... *I fear very much that it will not be found to concentrate in any great body of Ores – the veins are split up and distorted – nevertheless such indications may be thrown up from a large deposit underneath.*

In the morning, he called in at the Sherbrooke BALC office and was advised Galt had written to the mining company in London, ... *advising them to allow myself & 2 Men to come out next Spring to continue our exploration.* Galt judged the cost of such a venture was inconsequential, given the possible encouraging result that may come about, especially as Josiah was now more acquainted with the country and its terrain:

> *I am to give the Men notice that their services will not be required after the end of November.*

As this transaction had been left to him ... *I shall enter into an agreement with them to pay them their full time till arrival in Liverpool.*

The following Sunday, Josiah completed a letter to his father, James, as well as John Tregoning, indicating he would likely be home around mid-December. He also wrote a long letter to his wife Elizabeth saying how thrilled he was that she and the new baby boy, to be named Josiah, were both well after the birth on 10 October:

> *Attended English Church at Sherbrooke in the evening, as usual a collection.*

Josiah woke to a rainy morning and drove by wagon towards Dudswell, about 18 miles away, where he found his men examining an area of small creeks around Stoke Mountain. Snow had fallen all night and by midday was about six inches deep, which prevented any further exploration. When the snow had melted enough for them to leave Sutcliffe's farm, they all headed out in the wagon stopping seven miles later at the small timber milling township of Westbury. Josiah left them to explore the area and drove on to Sherbrooke in hope of seeing Cummins. Arriving late afternoon, he met Galt; Cummins had yet to arrive:

> *Could not decide anything about disposing of our Men either at Acton or elsewhere – at 8 p.m. Mr Cummins called on me at the Magog Hotel – he said that he saw Mr Galt at the Railing Depot here, and that I must go and examine some Copper reported to exist at Brompton Lake.*

As to the Upton proposition … Mr Cummins stated that the owner of the Land at Upton wanted £300/-/- currency for the Mineral Right or £600/-/- for Mineral & Land, shall see Mr Galt on the 31st to settle this business – for my part I would not purchase the Mineral right or land either before proving of the Mineral appear to be lasting in depth – and this can only be ascertained by making a trial of it by Miners.

In the morning, Josiah checked in at the BALC office, where £15 remittance was waiting for him. As the snow had mostly melted, he left by wagon at 9 o'clock and after driving around fallen trees from the heavy snowstorms, arrived at the head of Brompton Lake in the early afternoon and … *found a Man at the head of the Lake with a Canoe – started down River and got to an open log shanty at 4 p.m. In the evening made a fire on the beach and drew a nett catching 2 large trout & 10 herrings.* While the shanty was welcome, it did little to assist on a freezing night.

It was an uncomfortable night so Josiah rose early and left the shanty at 6.30 bound for Cathedrale Mountain on the southern side of the lake. From the shore, he saw two or three small patches of copper greens on the face of a vertical rock some 150 to 200 ft above and directed the guide to climb to the area where he broke away a number of yellow rock pieces about the size of a French nut: … *left at 1 o'clock and got to the head of the river in 1½ hours.*

Josiah paid and thanked the guide and returned to Sherbrooke through partially thawed muddy snow. With few opportunities for a more detailed investigation, partially due to the winter snow, his plans were uncertain … *what the Men will be put to do now I cannot say. But on Monday if the Acton affair is not arranged I shall recommend Mr Galt to send them to England directly.*

Galt arrived a few days later and told him BALC London had decided not to continue the Upton mineral lease proposition. Afterwards, Josiah gave the men notice they would leave Sherbrooke for New York enroute to England the next Wednesday, … *where they one and all appeared to be glad to return.* The following day the miners helped pack and store equipment that would be needed for next year's exploration. Excited to be returning home, they all met early at Sherbrooke Station and caught the 7 o'clock train to Montréal arriving at 11 o'clock. … *Mr Galt went by same car to Montreal and I had a long conversation with him relative to future operations in the event of the Company sending us out in the Spring of 1854 – remained at Montreal till next morning – purchased Moccasins.*

At Longueuil Station purchased rail tickets for the 250-mile journey to Burlington on Lake Champlain, then to Whitehall and onto Troy where they would catch a boat down the Hudson River to New York. Unfortunately, they missed the boat but found accommodation close by and reached New York by rail in the morning. The 150 mile trip took just over four hours. From the station in New York, they caught a carriage to the Glasgow Steam Packet office only to find there were no steerage berths available. Second-class berths cost $50 and after some discussion with the men, it was agreed they would receive $55 each, thus allowing them to make their own way home to Cornwall as they chose. The men purchased tickets on the *Cornelius Grinnell* packet ship which was expected to leave for Liverpool during the ensuring week. Having secured transportation for the men, they all walked to William Robertson's, 62 Greenwich Street, New York, where they had arranged to stay the night.

After breakfast, Josiah left the men after thanking them for their assistance and wishing them well on their journey home, then walked along Vesey and Church Streets, just five minutes away to 9 Cortland Street and booked a room at the Western Hotel. The Western Hotel was one of the finest in New York offering a high standard of service with bellhops and attentive waiters. A spacious lobby led to an elegant restaurant and a fine bar. Leaving his baggage, he then caught a carriage to Crystal Palace, recently opened in Bryant Park on 42nd Street between Fifth and Sixth Avenue. The 50 or so foreign stalls were not as extensive as he had expected, … *nevertheless it gave me great satisfaction – the picture gallery excellent, and the interior of the Building was very tasty — about 10000 people admitted – refreshment rooms good.* Being Sunday, he wrote to wife Elizabeth informing her of his plans and the likely day he would arrive home to Chacewater.

Marmora

7 November to December 1853

After a quick breakfast, Josiah arranged for the bellhop to hail a cab to Grand Central New York Rail Station. At the ticketing office, he bought a $6.95 ticket for the 6 o'clock train to Cape Vincent via Troy and Rome, the almost 400-mile journey taking a full day. From the station, he walked to the wharf and caught a boat at 7 p. m., arriving at Kingston at 9 o'clock. He found accommodation at the North American Hotel, 77 Princess Street, near the docks, $1.50 per day from his own expenses. Snow had started falling late the previous day and by the morning the heavy drifts were now slowly being dispersed by rain. The cold weather was taking its toll, and he was feeling far from well, ... *very unwell with a Cold – spitting and discharging in my nostrils had I been at home would have laid in bed.* He rested until mid-afternoon and then caught the 4 o'clock steamer to Belleville. Unfortunately, a late evening storm had built up across Lake Ontario, forcing the steamer to take shelter in a nearby bay. By morning, the waters had calmed enough for the steamer to leave, stopping at the Moorings before reaching Belleville Wharf at 3 o'clock. Disembarking, he walked to his accommodation for the evening and then found a livery to arrange for a horse for the following day.

On a fine frosty morning, he left Belleville at 8 a.m. for the 30-mile journey on horseback to Marmora. Earlier rains had reduced the road to not much more than a track. While crossing a bridge over one of the many creeks, one of the planks on the bridge gave way and the horse's hoof slipped through. ... *The horse fell and with difficulty saved a broken leg to it – I was thrown, had my mind knocked out for a short time but received no other hurt.*

Dazed somewhat, but lucky not to be thrown into the icy waters, he recovered and walked for a while, then mounted and continued on through to the small village of Stirling, and then Marmora arriving mid-afternoon. He arranged to stayed at the St. James Hotel, 33 Forsyth Street, for two nights, then made arrangements to visit the Marmora Iron Mine the following morning. The hotel was impressive, offering 40 elegant guest rooms and stables for horses.

Leaving early, he rode to the Blairton iron mine site some 5 miles west on the shores of Lake Crowe and was immediately impressed by the scale and size of the open pit construction. Still unwell from his cold, he was able to find shelter from the cold and wind and spent some time writing notes which would form part of his Iron Mine report for the gentleman in Montreal who engaged him. He then returned to his hotel in Marmora. The next morning, he rode to the smelting works alongside the Marmora River with its many water wheels, some over 25 ft diameter. The paddles measuring more than 6 ft wide pumped two massive bellows. There were few people about, the township mainly deserted, the smelting works having been closed since 1849. He took notes while viewing the two well-constructed stone furnace houses before returning to Marmora just before midday. From the hotel, he collected his valise and rode to Belleville arriving at 7 p.m., and returned his horse to the livery. The next day, after a leisurely breakfast, he took a carriage to the Belleville dock and caught a boat at noon to Kingston arriving at 11 in the evening. Rising early, he boarded a steamer at 7 o'clock and headed up the St Lawrence River for 70 miles arriving at Prescott at noon. As Prescott was on the western side of the St Lawrence River, he found the steamer ferry boat that crossed to Ogdensburg. After a quick lunch at the stately St Lawrence Hotel on the corner of Ford and State Streets, he caught the 2 o'clock train to Montréal arriving at 10 that night. He had hoped to collect any letters for him at the BALC Office before he left Canada, but after much searching was unable to find the address of the office clerk who might open up the office. Disappointed, he boarded the 6 a.m. train to New York arriving at 9 p.m. and caught a cab to William Robinson's, 62 Greenwich Street. New York.

Returning to England in early December, he made his way to London and met with Devas to discuss his report. On 19 December at the BALC special general meeting held at the London Tavern Mr J J Cumming stated that:

Capt. Holman and six miners had proceeded to the estates of the company in June last in order to make an exploration for minerals. The search, on the whole, was not of a very satisfactory character, because although gold was found in various localities, and, also, they were not found in sufficient quantities to warrant extensive working. The survey, however, was of a very hasty character, and Capt. Holman had scarcely reached Montréal on his return home, when Mr Galt, the Commissioner, telegraphed for him to say that important information had been received of gold deposits and washings of a very great value. The message, however, was too late, as Capt. Holman had left Montréal. Mr Galt then dispatched the company's surveyor with some assistance to the place, and they had succeeded in procuring and sending home a box of specimens, which, when examined by competent persons in this country, were declared to contain gold of a rich quality, and very similar to that found in California. The specimens were sent a few days ago to Messrs Johnson and Mathey to be assayed, but their report had not as yet, been received.

The 1854 Expedition

28 February to 30 April

Josiah had a joyous reunion with family and friends at Christmas, 1853. It was the first time he had set eyes on and held his newborn son, Josiah Jnr, just two months old. He was grateful his father John and mother Grace had been able to support the family until his return. Celebrations over, he returned to work managing the Creegrawse mine and St. Michael Penkevil mine for Humphry Willyams. In his report of 20 January 1854, he wrote of the problems facing the St. Michael Penkevil mine operation:

"The lode in the adit level, driving east of St. Mount's shaft, on the north lode, was 1 ft wide, worth £5 per fathom for tin. The winze below the adit level, west of St. Mount's, on the north lode, was suspended on account of water; the stopes in the east and west ends of this winze, near the bottom of the adit level, were worth £12 per fathom each for tin. In the 8 fathom. level west of St. Mount's on the north lode, the lode was 2 ft. wide, worth £15 per fathom. The engineers were actively engaged in fixing the new winding and stamping engine. It was expected that before the next bi-monthly meeting they will return nearly 5 tons of tin."

Earlier, in September 1853, Alexander Galt had advocated that Josiah should return to Canada with two men in the spring of 1854. Josiah was keen to continue his search and when Galt made representation to the "Court" in London, an agreement was reached. Captain Holman travelled to London in late February 1854 to finalise arrangements. Unfortunately, his journal of 1854 ends at Sherbrooke while he was waiting for Skews and Treweek, the two miners chosen for the expedition. In summary, we learn that on 28 February, 1854, he sorrowfully farewelled the family once more and travelled to Truro to meet Henry Tregoning — who had been in touch with Thomas Devas. Josiah agreed with his friend to insist on a 12 month contract, not the six months proposed by Devas. He left by coach for Plymouth, promising to write, caught the evening train to London and met Devas to discuss the expedition and finalise payment:

A few days since I was informed by Mr. W. H. Tregoning that he had been applied to by Mr. Devas of London on what terms Cap. Holman would go on another expedition to Canada and Mr. Tregoning knowing my mind on this matter wrote an answer stating the terms to him. 500 per year clear of all expenses — one year's salary certain.

At the request of Devas, he inspected the Penkes gold crushing machines and later visited their works meeting up with the engineer of Fowey Consols, Mr W West, as well as Captain Brakinshire of the Brazils. Over dinner they discussed at length the machines capability:

These Gentlemen did not approve of the machine with myself – it has not the power to dispatch a large quantity of Stuff. Still for testing a sample it does good justice by it as it reduces it to a complete powder viz 6400 meshes to the sq. inch.

Josiah returned to London the following day and met Devas to report on the Penkes machine and discuss and gain agreement to take stamps to Canada.
He judged them, ... *the only method of our making a complete & satisfactory trial of the Auriferous Quartz.*
Josiah was also informed that Alexander Galt was expected in London and he should wait for his arrival. As no meetings had been scheduled the following day, he caught a bus to Sydenham and visited Crystal Palace:

Passed through all the Rooms — a fine collection of paintings was the only attraction -- the pleasure grounds are beautiful and the ponds abound in various coloured fish – returned by Rail.

While waiting for Galt, he arranged a meeting with Mr Brischoff who showed him Galt's report of the gold discovery at Lake St Francis. Late that afternoon, Devas advised Galt had been delayed, and to take the chance to return home, which Josiah did by rail and road. Fellow passengers on the bus from Plymouth to St. Austell included a Mrs Phillips, her sister and two children. Josiah had met Mrs Phillips' husband, John, in Montreal the previous summer.

The family was of course delighted by his temporary return. The next day, he visited his father on the way to Redruth and the Perran Foundry Co. at Perranarworthal. He ordered stamps which were promised to be completed by the 20th:

From 13th to 20 instant inclusive I was engaged as much as possible at Creegbrawse Mine, making preparations on my outfit and family arrangements during my absence. to Canada.

Having completed as much as he could at the Creegbrawse mine, he arranged to meet Jno Skews, Ben Paul and Captain Blight and travel to the Perran Foundry to inspect the stamps. While the stamps had been completed, they had not been weighed, so he instructed they be forwarded on the 28th. The following morning ... *went to the house of Mr. W. H. Tregoning thence to my Fathers and dined — afterwards went through the very unpleasant duty of bidding my dear Parents farewell — returned to my home at 5p.m. and then had to perform a similar unpleasant duty in bidding adieu to my dear wife and children – in bidding adieu to my parents, I have this impression in my mind that if I should be successful in Canada and remain there a few years in all probability I may never see them again on earth but I trust God we may meet in Heaven but with my wife and children should my stay in Canada be long I should send and get them to me. I left home at 8 p.m. & reached Truro, supped with my friend Mr. John Tregoning called on my sister.* Rising early, he caught the 4 a.m. coach from Truro arriving in London later that evening.

Next day he met Devas and was advised that a meeting with the mining company (BAMA) would be held on the 30th and that he was to attend. He also cleared up a misunderstanding on his remuneration:

Mr. Devas thought my payment was for 500 clear of travelling expenses only. I said it was clear of all expenses including Board and therefore to meet the misunderstanding Mr. Devas agreed to make it 500 Guineas and clear of travelling expenses only. He said that Mr. Pennoyer had much less salary.

Returned to the Portugal Hotel and in the morning received a note from Bischoff:

Met Mr. Bischoff this afternoon, he purchased a gun, Colts revolver and a mackintosh for my use.

Later, he attended the meeting of the Mining Co. at the BALC office, 24 New Broad Street, present Messrs. Bischoff, Devas, Cummins, Galt and two other Gentlemen. Talked over the various business topics connected with future Mining operations in Canada. Mr Bischoff promised to send by following Steamer written instructions to Mr. Pennoyer.

Before leaving for Liverpool in the evening, Josiah caught up with his elder brother, John, who had returned from Norway and was looking quite well. In Liverpool the next morning, he caught up with Skews and Treweek and walked to the Steam Packet Office where he was able to secure a berth on the *America* bound for Boston. Unfortunately, he could not purchase a berth for Skews and Treweek on the *America*, but instead the *Cleopatra*, £12/12/-, leaving 10 days later.

On 1 April, he boarded the 280 ft wooden hull steam-sail ship *America* with a full cargo, and reached Halifax on 14 April, after a squally, rough voyage passing many icebergs. Here are his notes on the trip:

6th. *Ladies all made their appearance on deck for the first time since the day or sailing — all parties getting in good spirits — have a good appetite but bad headache.*

7th. *... played Shuffles — headache continuing — all sorts of games carrying on by the passengers. Ladies singing.*

10th. *A very rough wet morning — heavy side swell causing a great commotion with the breakfast things — took breakfast and went below & vomited several times lay in bed all day, eat a little at dinner time — towards evening the wind & sea got calmed — several icebergs close by at evening.*

11th. *Fine, clear cold morning, got up at 7 a.m. saw several Icebergs — one passed within a 1/2 mile - was 30 ft high & covered several acres — it had hills & valleys & the top covered in snow having the resemblance of a barren Ocean Island — after 9 a.m. saw no more Ice — towards night a dense fog — which was anything but comfortable in the vicinity of Icebergs.*

12th. *Side wind heavy & rough & wet — felt anything but well — did not vomit — looking earnestly for the end of the voyage. I dislike the Sea and have suffered more from Sea sickness than I expected. Every kind of food disagrees & am very bilious. The steamer is very slow owing some say to bad Boilers others to bad Coals.*

Arriving at Halifax at 2 a.m., Josiah went ashore after an early breakfast and wandered around the township: *... the town is pretty large but the majority of Houses are built of wood. Several churches — the town is on the side of a Hill, the summit of which is crowned by a strong fortress.* Snow was falling heavily when they left at 11 a.m. the following morning for Boston, although the seas were calm. Arrived in Boston late the following afternoon and after clearing customs the following morning, caught a cab to the Revere Hotel. Being Sunday, he attending a Unitarian divine service in the afternoon:

Heard an eloquent sermon, but I observed with regret that the congregation sat in the same position at prayer time as at sermon. Did not see a dress coat in Boston save the one I wore. I was much pleased with the buildings, they are very substantial and ornamental, no expense appeared to be spared to arrive at these desirable ends. I think I never saw finer blocks of Granite than is frequently to be met, with these houses and buildings. The Revere Hotel is the best House of the kind I ever put up at, the Yankees go ahead of the English in this aspect.

He rose early and caught the 6 a.m. train to Island Pond via Portland — a trip of nearly 300 miles, finishing in the early evening:

At Island Pond my baggage was overhauled on entering Canada — however the officer knowing me from last season he was satisfied by my informing him that I had nothing in my baggage but for my personal use.

Leaving Island Pond at 6 a.m., arrived at Sherbrooke at 8 a.m. and collected £25 from the BALC office — final payment from Galt for inspecting the Marmora Iron Works the previous year. Josiah met Pennoyer, who had returned from Lake St. Francis: *... spent the evening at his house with Col. Moore and heard most of the circumstances connected with the gold discovery at Lake St. Francis and that the same is secured for the B.A. Mining Co.* In the morning, inspected a specimen of gold which Pennoyer had obtained from some of the quartz found on the surface at Lake St Francis:

Informed Mr. P. yesterday that his instructions as well as my own were promised by Mr. Bischoff the day I left London to be forwarded by the next Steamer. Mr. P. was disappointed on having no letter sent from the Company by me. I told Mr. P that our both positions would be clearly defined on the Instructions which are to be forwarded.

Leaving Pennoyer, he walked to the nearby foundry and ordered an iron grating for separating the stones from gravel when gold washing.

The following day, he and Pennoyer met briefly and discussed how best to get to Quebec City and later had tea at the Andersons:

All parties here say that I look much thinner than when I left this place last fall — this I attribute to having suffered much sea sickness.

Left Sherbrooke by stage in the early morning, reached Cookshire at 1 o'clock, secured an untried horse and small wagon and drove to Bury. *Very little snow on the road or land except in the drifts. The Ice is thawing on the road making it very soft and almost impassable for carriages.* The next morning, he tried riding the horse to control it better:

It is an unbroken obstinate colt – it shied into a deep ditch of Snow & threw me of but did but little injury.

He arrived at Gould at 11 o'clock, having covered only 12 miles in nearly five hours. In exasperation, he returned the horse having ... *paid an unusually high sum (40/Cy) for the trip.*[*] He then hired a two-horse sleigh, hoping it would be more comfortable, but got thrown out of it before even reaching the first river. On reaching the river, he was told the ice was not broken, so he sent the horse and sleigh back and shot two partridges while waiting to organise a crossing.

Leaving his baggage, he crossed the river and searched for another horse and sleigh. After finding one two miles away, he backtracked for his gear, then had more strife while chatting with the rig's Irish owner. Suddenly, the new horse started off: ... *near the house the sleigh was knocked to pieces but did not much injury to my trunks.* He stayed overnight with the Irishman, travelled all day to the Felton River at the head of Lake St Francis and reached St Francis at noon the next day. Despite the danger, he was able to cross the Chaudiere River on the ice and arranged for a sleigh to take him to Mr Calway's. Heavy snow continued all day. Standing in 12 to 15 inches of snow, ... *the cattle about*

---

[*]   Cy is Canadian currency, 40 Cy being 40 Canadian shillings.

are starving for want of food — the thatch or straw coverings on most of the Barns are being taken up for fodder for the cattle.

There was no news in the *Quebec Gazette* of the *Cleopatra's* arrival. So, Josiah had no option but to wait and wrote to George Pemberton asking him to make contact with Skews and Treweek when they arrived. He left Calways early afternoon the following day, recrossing the Chaudiere in a canoe and returned to Tring at 6 p.m.:

This is the first instance of my setting out on a journey to reach a place in short, attaining my object. A little raining all day, however was better sleighing than I expected.

Sunday – Heavy rain last night and continued all day rendering it very disagreeable to take a ten hour journey hence I remained here — in fact the people here say it is almost a matter of impossibility to cross the valleys today as the rivers will be unusually high — and I do think it right to be breaking the commandment in travelling on the Sabbath but had it been fine weather I am afraid the temptation would be irresistible.

It is unfortunate the journal concludes at this date.

When the St Lawrence finally thawed in May, Josiah would have met up with Skews and Treweek, hopefully at Lake St Francis. In many respects, the expedition over the next six months would have been similar to the previous year, concentrating on exploring the new areas identified by Galt. Josiah made many other references to his visit to the native copper mines of Lake Superior, beyond his January 1859 letter to John Christoe:

In 1854 1 visited several of the most celebrated native copper mines on the southern shores of Lake Superior, United States of America.

It is likely this visit would have been towards the end of 1854 on his way to New York to return home. Although copper had been discovered many years previously, commercial mining had only begun a few years earlier on the Keweenaw Peninsula, Michigan. The winter closing in, Josiah, Skews and Treweek would have returned to England in November or December, Josiah then making his way to London to present his report.

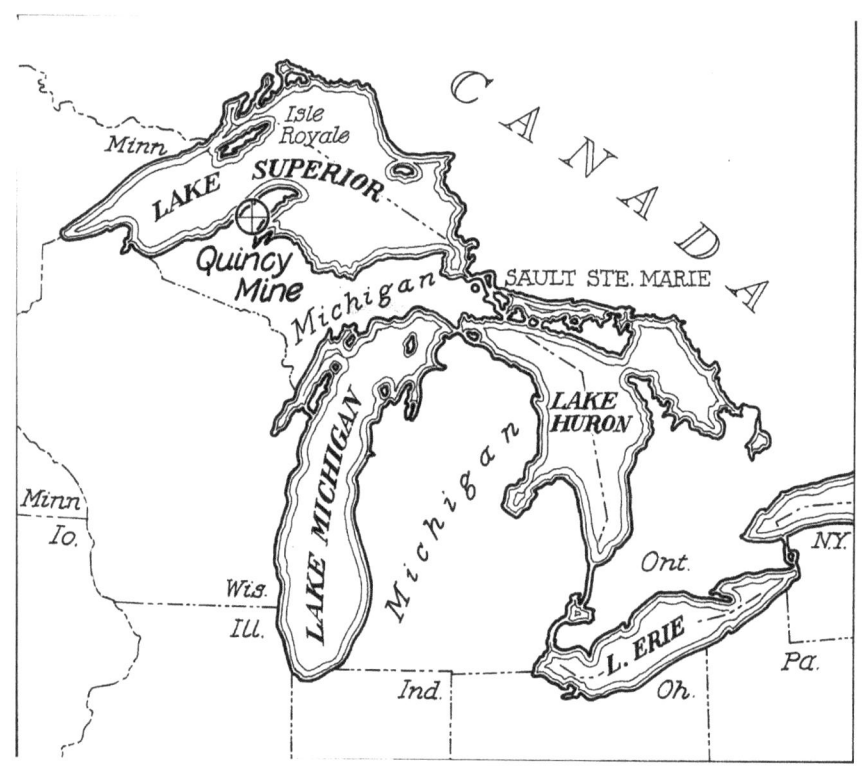

Library of Congress, HAER Record

PART THREE
SOUTH AFRICA, MALACCA STRAITS & NEW ZEALAND
1855-1862

South Africa

1855–57

Exploitation of mines in South Africa had been carried on unabated for years from around the 1840s until mining legislation was introduced in the early 1850s. In fact, for centuries indigenous Bushmen had been fashioning iron for spears and copper for ornamentation. Early explorers had observed mining during exploration of the interior, some taking notes of areas where the natives had carried out limited excavation. Trials had been carried out around Springbokfontein (later Springbok) as early as 1685 and later in 1779 with limited success. Renowned geologist A. G. Bain had conducted a number of geological surveys throughout the region and reported on the mineral composition and topography. In early 1854, Bain was asked by the Governor to carry out a geological survey of the Namaqualand district. Collecting samples from various areas, he tabled his report to both Houses of Parliament in Cape Town on 25 July 1854 noting that yields were generally around 45 percent copper. Others such as Atherstone, Bell, Damant, Wyley, Moffat had written voluminous reports and mapped some of the Namaqualand mineralogical areas. One of the first to recognise, or at least give consideration to the possibility that maybe, just maybe, there could be riches to be gained was the firm of Messrs Phillips and King. Established in Cape Town and described as a trading and mercantile house, they would later become involved in mining, one of a few that tasted lasting success.

In March 1850, Phillips and King entered into a questionable arrangement to purchase large tracts of mineral property in Namaqualand, comprising 202,940 acres from the Cloete family for £2334/10 -. Prospectors and geologists had begun exploring Namaqualand, around the villages of Okiep, Concordia and Nababeep, and reported that copper veins could be seen exposed on the barren hillsides. Through the early flurry of copper mining, from 1852 to 1872, Hondeklip Bay acted as the main port for the Namaqualand mines, with ore transported to the coast by road through the Messelpad Pass. As is often the case, small discoveries very easily escalate into a "rush to riches" and so it was in South Africa in the early 1850s, newspapers fanning the flames of fortune:

> **13 September 1854**
>
> The field of mineral wealth opening up in Namaqualand appears to be immense, and seems likely to surpass anything else in the world. In that portion of the Colony there are copper mines of almost unparalleled richness. The ores produce, on the average, from forty to forty-five percent, of copper; and in some cases, as high a per-centage as seventy-five percent. has been obtained. The ore appears to be obtained with the greatest ease, the mines, in most instances, having been hitherto worked as open quarries; and one miner and two assistants are able to extract and prepare for shipment about one ton per day.

By the mid-1850s, there were more than 30 mining companies registered, each promising to repay shareholders with fabulous returns. Phillips and King's Cape of Good Hope Mining Company (CGHMC) was one of these. In August 1852, 11 tons of copper ore was sent by ox-wagon to Hondeklip Bay to be shipped via the steamer *Bosphorus* to Swansea, Wales, and when smelted produced pure copper averaging 29.7 percent. On the back of this impressive result, the company issued a prospectus in November 1853 (Deed of Settlement 18 Jan 1854) to raise capital to purchase and explore further their leases by offering 10,000 shares at £10 each. Their leases were extensive and they had already begun limited exploration of the Springbokfontein, Spektakel, Nababeep and Okiep mines. By early March 1854, shares of the CGHMC (at this stage 10/- per share paid up) had soared, merely on speculation, to £4 8s each. Initially, small quantities of high-grade ore were shipped out from the Namaqualand mines and by the end of 1854, more than 1000 tons of ore had been dispatched, almost 150 tons from Phillips and Kings Wheal Julia mine. In late 1855, some 200 tons of copper was shipped from Hondeklip Bay on the barque *Dido* for Swansea with a further larger quantity awaiting loading on the brig *St Vincent*. It was reported that some ingots contained 75 percent pure copper. Fuelled by such reports, leases of little value, often sight unseen, swapped hands on a weekly basis, the value spiralling with each exchange. It was reported almost one and a half million Sterling had been nominally subscribed.

COPPER MINING SHARE LIST.
Cape Town, Dec. 12, 1855.

Name.	Established	No. of Shares	Nominal Value £	Amount paid up. £ s. d.	Present Price cash. £ s. d.
South African	846	2000	10	3 0 0	22 0 0
Cape of Good Hope	853	10000	10	1 0 0	5 0 0
Port Elizabeth	1854	10000	10	0 10 0	0 15 0
*New Burra Burra		5000	50	0 5 0	3 0 0
Walwich Bay		2000	20	1 0 0	20 0 0
*Orange River		1000	25	5 0 0	12 0 0
*Spectacle		300	50	5 0 0	6 0 0
*Grt. Namaqualand		2400	100	1 10 0	50 0 0
Grah'sTown Great Namaqualand		2500	5	0 2 6	0 8 0
Western Province		300	5	5 0 0	17 0 0
The Paarl		7000	10	0 2 0	1 0 0
Paarl Kolbe		6000	10	0 10 6	5 0 0
*Nabas		3000	10	10 0 0	12 0 0
Equitable		10000	5	0 5 0	0 7 6
*Union		10000	10	0 10 0	0 12 0
*No. 6		4800	5	0 10 0	0 11 0
*Alliance		12000	10	0 5 0	0 6 6
*Cape Colonial		6500	5	0 10 0	0 10 0
*Victoria		6 00	5	0 10 0	0 11 0
New Great Namaqualand		500	5	0 10 0	0 11 0
*Eagle		5000	5	0 10 0	0 10 0
Tradesmens		10000	5	0 2 6	0 3 0
*Namaqua		10000	20	5 0 0	} to be paid up.
*Hare's		3000	12	5 0 0	
*Henkries		4000	10	0 10 0	

* Those marked with an asterisk are offshoots from private Companies, in which nearly one-half of the Shares, or a greater number in some instances, are retained by the original proprietors.

Sir George Grey was appointed Lieutenant Governor of South Africa and when he arrived on 5 December 1854, was confronted with this calamity. It is unlikely there was a better, more experienced administrator in the Empire to resolve such complex and challenging colonial legislative issues. His tenure as Governor in South Australia (1841–45) was marked by significant reforms and advancements that helped stabilise and develop the region. As Governor of New Zealand (1845–53) he set about addressing the delicate political and social dynamics, implementing policies that effectively balanced the interests of both European settlers and the indigenous Māori population. His diplomatic skill and reformative approach earned him widespread respect and established him as an exceptional leader. Recommendations were suggested as to how to contain ruthless profiteering, but unfortunately Grey failed to act on the advice proposed. Instead, he introduced new legislation for mineral leasing and on February 23 1855 the Cape Government printed in the *Government Gazette* the following:

> With reference to the Government Notice of the 13th September 1853, relative to Minerals on Government Lands. It is hereby notified that no further applications for the lease of such Mineral Lands, under the said notice will be received after the 23rd April next.

The directors of CGHMC were not about to be swept up by all this hysteria, they had reputations to keep and they needed to respond to the market's request for information. The *Cape Town Monitor* of 20 January 1854 commented: "Shareholders are becoming anxious, and in many instances the uncertainty is acting injuriously to companies."

Newspaper reports from Australia had shed a light on the fragilities of a mining boom and carried stories as much about the riches, a few, as it did the losses, many. Just as in Australia, naive understanding of the countryside and its environs saw many mining companies fail, so too Namaqualand where little exploration had taken place. Earlier, Lt.-Governor, C.H. Darling commented upon Namaqualand:

Of the particular district which is thus about to become the scene of mining enterprises far less is known, with certainty, than of any other portion of the Colony ...

With Grey's new legislation preventing any further purchase of leases, rich or not, mining companies had to rely on their current portfolio, of which there were conflicting reports. Geologist A G Bain wrote of his 1854 expedition:

I next visited the mines of Prince, Collison, Watson, & Co., at a place called "Tweefontein" now called Concordia. These are also exceedingly rich, but unlike the other mines I had visited; has something of the appearance of immense lodes running in an Easterly and Westerly direction, yet from their length, breadth, and depth, seemed inexhaustible. Here I saw some of the richest and most beautiful specimens of copper pyrites peacock, and malachite, etc., that I had anywhere met with. For miles around this favoured spot, strong indications of copper everywhere appear; and a large village is in the course of construction in which hundreds of happy families may yet reside. Much activity prevails here, as well as at Phillips and King's mines, but more in the way of building than of mining, as transport to the coast is not at present to be obtained, even at very high prices.

When Josiah returned from his second expedition to Canada in late 1854, the family moved five miles to River Street, Truro, Cornwall. He continued managing the Creegbrawse and Penkevil mines for Humphry Willyams but recent reports indicated the rich lodes of the past were quickly declining. Under Captain John Blight, who was appointed in 1853, much had been achieved while he was away. Reading the most recent reports and after an examination of the shafts and ore, it became apparent the mine was unlikely to continue operating profitably for much longer. In early 1855, Josiah was asked if he would be interested in carrying out an assessment in South Africa, similar to the one in Canada. It is most likely the approach was made by John Taylor & Sons on behalf of the Cape of Good Hope Mining Company.

Taylor had made similar offers to Josiah in the past. After agreeing terms, he resigned as manager of the Creegbrawse and Penkevil mines, leaving the managing to Captain Blight. In April 1855, a meeting of Creegbrawse mine shareholders was held and it was agreed to wind up the affairs of the company. In July, an advertisement appeared offering for sale a pumping engine, crushing machine, stamping and drawing machine and in September a similar advertisement auctioning all machinery and parts for a mining operation to be held at the Creegbrawse mine, Parish of Kenwyn.

Accepting the offer to lead the expedition, his first task was to select a team of Cornish miners and begin planning for the expedition: travel arrangements, transportation, tents, mining equipment, food and water. The remoteness of the region was well documented; it would be an even more difficult expedition than the Lower Canada venture. Andrew Wyley, a geologist who visited in 1856 wrote, "Hot, dry and barren ... hopelessly so." Josiah's wife Elizabeth, having stayed at home during the Canada expeditions, was keen to accompany him. The children would continue schooling and would be looked after by Josiah's father James and his mother Grace. He also wished to gain more assaying experience and contacted John Penrose Snr, of Lemon Street, Truro who put him in touch with his son William Henry Christoe, a renowned Welsh Master Assayer. His initial task once he arrived was to provide a preliminary report for the special general meeting of the shareholders of the Cape of Good Hope Mining Company.

With his party of Cornish miners, mules, wagons, mining equipment and supplies, they arrived in Hondeklip Bay some 350 miles north of Cape Town around mid 1855. The bay was not deep, quite shallow in fact, which may account for the long jetty, over 150 yards jutting like an obelisk into the bay. A rail track ran the whole length of the jetty which was used for loading ore from the carts. Along the single street there were half a dozen or so trading stores, a Court/Customs House and from the beachfront cottages were scattered about the low hillside.

After a few days, and having purchasing fresh provisions, they waited as the brig entered the bay and docked at high tide at the end of the jetty.

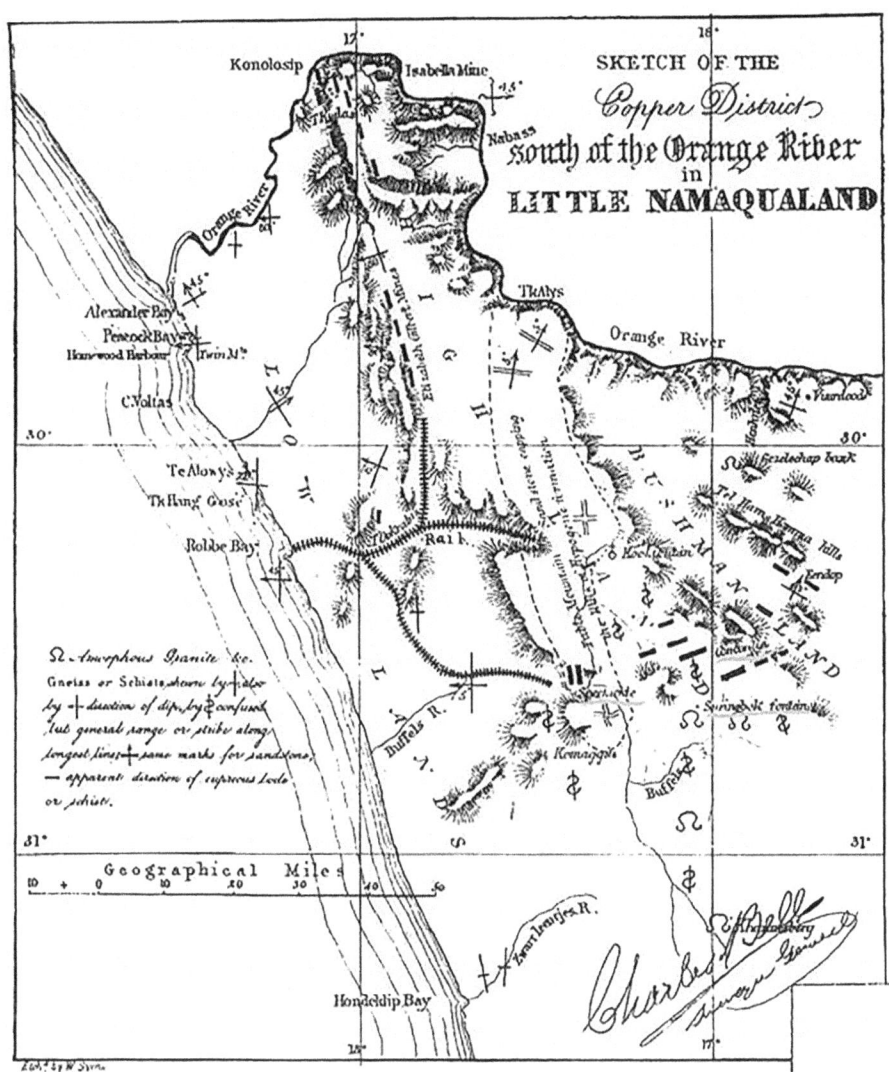

*Bells 1855 map of Okiep (Namaqualand) copper district.
Map showing the relationship of various Bays and Harbours to the mining districts
and the line of railway which Charles Bell proposed in his report of 1855*

After loading all their equipment, they headed north 100 miles to Port Nollath where they were able to secure more supplies and some additional equipment. Josiah had earlier contacted Phillips and King and arranged to meet entrepreneur John Owen Smith. Smith who owned the Kodas (Kuboes) and Schaap River mines also had shipping interests and was transporting ore down the Orange River via shallow draft barges, each holding 40 tons of ore using a 10 horse power pinnacle (small boat). Travelling this way would save four or five arduous days by wagon. They were to meet Smith at Port Nollath and after a day waiting, received information he was held up and could they make their way to Alexander Bay at the mouth of the Orange River some 60 miles further north. A few days later they were all able to obtain berths on a boat travelling north and at Alexander Bay, transferred all their equipment onto one of Smith's barges. Leaving at dawn they slowly made their way up the Orange River flanked by shading trees, drooping willows, except when engulfed by gorges many hundreds of feet high. It was the only respite from the heat and the following morning they reached their destination a few miles from the Numies (Numees) mine.

Unloading their equipment onto mule wagons, they headed east to the Numies mine, a largish open pit with a stope extending horizontally into the mountainside and ventilation shaft from the top of the rise. The men were instructed to excavate shallow pits further along the mountain side as well as the other side of the valley in the hope of discovering new lodes. There was initial excitement as the men unearthed clumps of rich copper embedded in the quartz strings, but these did not "trace" any further. After a few weeks, and now having become more accustomed with the climate and to a certain extent the barren terrain, they packed up tents and equipment and headed south five miles to Kaboose (Kuboes). At Kaboose, they established camp and began exploring the area where some diggings had already begun. Ore had been taken out at the Jessie Smith Mine, shafts having penetrated 24ft and 60 ft. There were also a number of superficial minerals exposed in the area to keep the men busy, but the discovery of any deep rich vein proved elusive. Occasionally they would uncover exquisite pieces of crystallised quartz of varying brightness and colours, but the reported copper ore of 45 to 75 percent was nowhere to be found. Here, Elizabeth was able to obtain fresh food and establish a camp for the next few weeks.

After almost a month and not having exposed any mineral lodes of note, they loaded up the bullock-drawn carts and with the native guides headed south from Kaboose. Crossing the barren rocky mountain desert Bush and Red

Sandy Flats, they continuing over desolate rugged terrain until they reached the main road from Port Nolloth to SteinKopf (Steinkopf) near Oograbees. After a few more days, they reached Buffells River, having completed the arduous 150 mile journey in just over a week. At times, the weathered granite mountain range gave way to dome-shaped hills and in between huge granite boulders often boarded shallow valleys. A few miles to the east was Owen Smiths Schaap Rivier mine and four miles south from there, Phillips and Kings Spektakel mine. Josiah had promised Smith he would spend a few days at his mine examining recent excavations. Below rocky cliffs, a shaft about 15 ft had been tunnelled in the hope of finding a copper lode. It was only around the entrance that copper ore was found. On the surface, the granite rock contained quartz and mica with occasional fractures of copper, but proved difficult to break, so Josiah instructed the miners to focus a little further away from that area digging shallow excavations where they could. A village, if it could be called such, consisted mainly of mud huts and tents would test the most adventurous of prospectors.

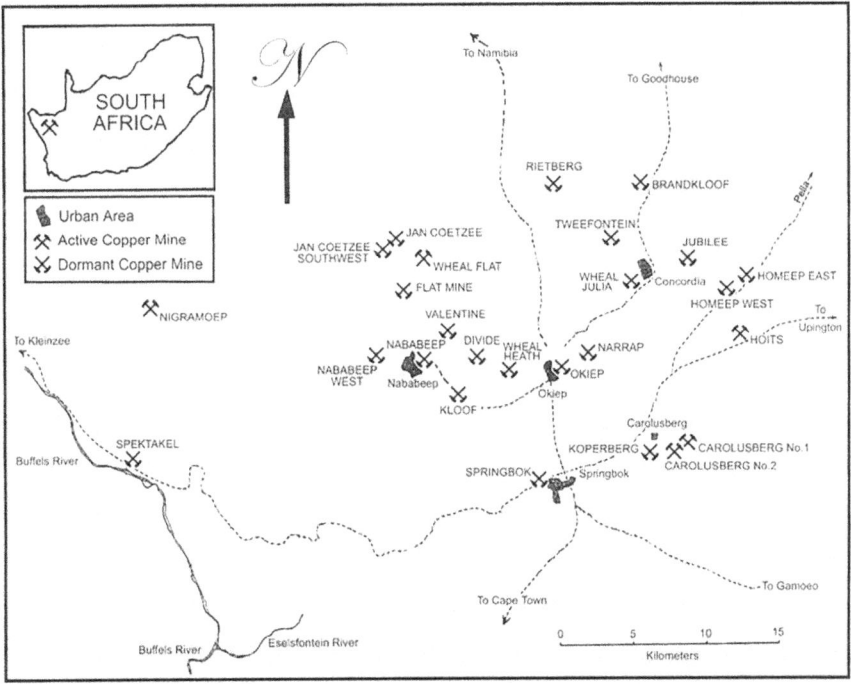

Okiep Copper district map – circa 2000

At Spekatakel, Josiah and the team examined the quarry that had been carved out to a depth of over 20 ft. CGHMC had commenced mining the previous year and begun excavating three pits in an open cut manner with some success. Entering the pit, Josiah examined the green carbonate mineral in what could be described as "clumps" of Chalcopyrite intermixed with brownish blue copper ore. Piled high surrounding a number of small mines were heaps of green stained ore awaiting shipment to Hondeklip Bay once cartage costs made mining viable. Speaking with a number of the industrious miners, it was planned to blast deeper into the quarry, the miners not yet convinced they had reached the end of the lode. They assured Josiah the yield was about 45 percent copper, but after lengthy examination of the ore heaps, he concluded the yield to be more likely half that. However, his miners did on occasion find "clumps" of ore close to the surface, some of which he determined were around 70 percent pure copper. It was both unusual and exciting for the miners to find such rich lodes so close to the surface, their Cornwall experiences relied on deep digging before a reef was located. Finding a sizable reef of such rich ore would certainly secure the viability of the mine. Josiah selected a number of areas where the rock was not as hard and conducted a trial for a month. Having excavated only two tons of copper ore and not finding any true lodes or continuous veins, he determined future prospects seemed discouraging, so their works were abandoned.

From here they continued east to Kings and Messrs. Prince, Collison and Company's mine, Hester Maria in the Concordia region. A Cornish community had been established some years earlier and was beginning to attract more miners seeking their fortune. Development and building were active, the streets well planned and laid out, a testament to the faith the miners had in this area. For Elizabeth, the village "atmosphere" was welcomed as she was able to secure fresh meat and vegetables plus a much-needed bathing opportunity. It was at this time she realised she was pregnant with her fifth child. Here the ore ran in reefs, as it did in many parts of the world Josiah had visited, contained very few impurities, the ore having been assayed up to 33 percent copper. One such splendid lode, 16 ft wide, had been traced for a mile in length producing 23 to 26 percent pure copper when shipped to Swansea in May 1855. This was by far the most impressive Josiah had seen since arriving in Namaqualand. However, lack of timber for fuel would mean ore would need to be carted to the coast and shipped for smelting, a costly undertaking.

Josiah had been requested to complete a preliminary report to the CGHMC special meeting of shareholders to be held in December. His years of experience managing the Creegbrawse and Penkevil mines had taught him that discovery of rich ore was only one element of many that determined the success, or not, of a mining venture. All other considerations would need to be aligned before he would recommend continuing operations and the outlay of additional capital. Now, almost the end of November, he spent the next few days completing the report while the miners continued exploring various outcrops of quartz in the hope of finding copper reefs. At the meeting held on 15 December 1855, Holman's report was tabled and it was reported that at ....

"A special general meeting of the shareholders in the Cape of Good Hope Mining Company was held on Dec. 15, for the purpose of considering a report from the directors, founded upon information received from Capt. Holman, the manager, recently engaged with a party of miners from England. Capt. Holman had inspected the whole of the centres belonging to the company (excepting those in the locality of Concordia), and found nothing in them that justified him in recommending further operations. He, therefore, advised that they should be abandoned, but recommended the retention of the company's second, on Messrs. Prince, Collison, and Company's mine—Hester Maria; and considered that several centres, in the neighbourhood of Concordia, require more careful examination than he had hitherto been able to make. "

> MINING IN THE CAPE OF GOOD HOPE.—We have received intelligence from the Cape of Good Hope to Dec. 17. A special general meeting of the shareholders in the Cape of Good Hope Mining Company was held on Dec. 15, for the purpose of considering a report from the directors, founded upon information received from Capt. Holman, the manager, recently engaged with a party of miners from England. Capt. Holman had inspected the whole of the centres belonging to the company (excepting those in the locality of Concordia), and found nothing in them that justified him in recommending further operations. He, therefore, advised that they should be abandoned, but recommended the retention of the company's second, on Messrs. Prince, Collison, and Company's mine—Hester Maria; and considered that several centres, in the neighbourhood of Concordia, require more careful examination than he had hitherto been able to make.

Mining Journal, 15 March 1856

Josiah and the team of miners spent the next few months examining several centres in the neighbourhood of Concordia, Wheal Julia and nearby Phillips and Kings Kilduncan and Nababeep leases. These were examined in more detail, reviewing the area and prospecting mineral outcrops he had identified. Although there had been mineral prospecting of the area with a number of narrow stopes and open trenches where ore had been unearthed, mining had not yet begun in a significant way. They were able to extract some ore of very good quality, but in many instances, any trace of a rich lode soon petered out. At times, minute traces of gold were observed shining through the small cracks of extracted ore. There were however a number of shafts which had been sunk following the red ore (oxide of iron and copper) mineralisation which continued in an east-west direction. The specimens obtained from the western end very much impressing Josiah. With heavier equipment, which they did not have, he would have liked to explore deeper into the granite rock. Beginnings of a mining settlement were under construction, tents and huts taking up the most valued positions. At the company's Kilduncan mine, where high yielding ore had been extracted earlier in the year, they sunk some shafts but without heavy equipment, they did not penetrate to any great depth.

Nearby, five miles to the west, they also spent some time at Okiep prospecting around the beginnings of an open pit mine situated in the middle of the barren valley surrounded by irregular low mountain ranges. Much of the excavation had been undertaken using gunpowder to break the granite stone; a common practice in the Namaqualand mines. It was reported during April, 1855, almost 2500 pounds of gunpowder was shipped to Namaqualand. The men focused their efforts working an area where a shaft had been sunk, which continued over 20 ft following the lode, although the irregular veins varied in quality. Here they excavated shallow hollows of the decomposed genesis with green copper stain and red ore they found displayed at the surface. The miners were excited by their success. Nearby, miners, carters and labourers, coloured and white, were actively engaged working leases with pick and shovel, wheelbarrows, bagging ore, sewing up bags for loading and unloading carts. They seemed better equipped to handle the heat, some days up to 130 Fahrenheit. The Europeans, an eclectic mix of trades, artisans and farm labourers in the main had little or no idea of mining, their application to exploration clearly shown in their poor results. Piles of ore dotted the landscape waiting to be assayed or to be loaded into bags and onto carts.

Continuing on about three miles was another mountain rising to about 80 ft, Koper Berg, the blue green copper-stained granite appearing to be a mass of copper; malachite, blue carbonate of copper and occasional minute particles of gold. Numerous shafts of various sizes had been excavated into the base of the slope. Josiah went about prospecting the shallow exposed ore, directing the men to sink shallow shafts in the hope of finding a sizable vein of quality in the area. He also spent some time at the Cornetite Blue mine below a rocky hillside. The men were particularly impressed by the abundance of rich blue ore, something they had not sighted before. Having spent over six months examining the operations thoroughly, he concluded that without any new rich vein to excavate, further expenditure was not warranted.

Sketch of Springbokfontein, circa 1854. Courtsy Library of Parliameny, Cape Town

Travelling a few miles south, they reached Springbokfontein, pleased to resupply vital supplies from the local stores, get equipment fixed at the workshop and with Elizabeth attended the small church. Owned by Phillips and King, the township was showing signs of progress with roadways clearly defined and a number of houses completed or in the throes of construction as well as mud huts scattered about. There was a reasonably well stocked store, shops, store rooms, a post office, stables and blacksmith. Josiah spent the week setting up camp and directing the men to begin shallow excavations in and around the mine area. Within the valley of the Koperberge Mountains, chalcopyrite with black copper ore was intermixed with an abundance of superficial native and red copper ore exposed by centuries of wind and sand storms weathering the landscape, some of extremely high quality.

After three months mining the bare and desolate hillside and having excavated only two tons of copper ore of commercial quality, he decided to halt any further work. There was no section offering sufficient encouragement to warrant recommending the company begin a deep trial. He concluded that … *after the extensive research by many before him that Namaqualand has had applied to it during the past two years and a half without a single prize being obtained, leaves but little encouragement for the future.*

Later it was reported that in Namaqualand, accounts were very favourable and in one lease they discovered a vein of pure copper 14 ft wide, running the entire length of the shaft. Eight men were daily extracting over 14 tons of the finest copper, reportedly worth £75 per ton.

Having satisfied himself that no further exploration would reveal a rich lode, they packed up their equipment, tents and supplies and returned to the coast. The rugged barren terrain convincing him further just how difficult it was going to be to transport huge quantities of ore cheaply to the coast. Port Nolloth in Robben Bay was closer and the route not as hazardous as the Hondeklip Bay route, even so, expensive improvement was required. Cost had increased steadily to over £10 per ton cartage to the coast and return whether full or not. In fact, the cost to cart ore to the port was almost the same as the combined cost to mine the ore and ship it to England. It would certainly be a logistical challenge to maintain profitability if the current expensive mode of transport persisted. Surveyor General Charles Bell had surveyed the area and in 1854 the Namaqualand Railroad and Tramway Company was founded with a capital of £200,000. However, the magnitude of the task proved to be far greater than expected, capital was quickly exhausted and the project failed. A train line would certainly be of great benefit as well as a smelting operation in Namaqualand to reduce the ore to copper ingots. Geological Surveyor Andrew Wyley had returned from his survey and his extensive report had been tabled. The Cape Town Mail on 17 January 1856 commented, "He has before him the ancient Dutch surveys of these copper mountains; occasional notices by travellers, such as Barrow and Alexander; the more recent reports, founded on personal observation, of Bain, Beal, Holman, and Cameron..."

After almost 18 months and having completed the expedition and assessment of the company's mines in and around Namaqualand, in November 1856 Josiah caught the fortnightly steamer from Port Nolloth to Cape Town, his mining team returned to Cornwall. On 15 December, the Cape of Good Hope Mining Company held a special general meeting to consider the new report presented by Captain Holman. In part, it stated that after carefully surveying and testing the area he found no compelling justification to recommend further operations. He therefore advised that the area be abandoned, but recommended the retention of the company's second selection, on Messrs. Prince, Collison, and Company's mine, Hester Maria. He also recommended that several centres in the neighbourhood of Concordia should be examined more thoroughly. He wrote in his report that ... *None of the centres (ie the Concordia district) afford any appearance of their containing superficial deposits of ore at all comparable in quantity and quality with the productive mines in the district.*

To fully consider the report and other information from the area, the meeting was adjourned for a fortnight. At the postponed meeting, chaired by Dr James Abercrombie, Captain Holman's report was reviewed and discussed at length, especially in light of a proposal put forward that the company should sink deeper shafts to further test the existence of copper ore in that area. After a lengthy robust discussion, the directors not willing to abandon their mining leases, recommended that mining operations should be continued for another year, but limited to a cost of £2500.

Elizabeth, heavily pregnant and Josiah boarded their ship after these meetings and only having travelled a short way on their journey home, it is recorded on 13 January 1857 Charles William was born at sea off Cape Town.

Malacca Straits

1857-58

What do we know of Josiah's next assignment in Malacca? Unfortunately, from research, there is very little information, so let us start with what is known. On the 11 January, 1859, he wrote to John Christoe: *"Mr William Henry Christoe (elder brother) informed me about 3 months since, on my return from Singapore."* This would confirm he returned sometime in September 1858 and, taking into account the almost two-month journey, that would suggest he left Malacca and Singapore around July/August 1858. We certainly know he had returned sometime before the end of September, as on 30 September he placed an advertisement in the *West Britton* newspaper advising:

> I hereby give notice that I will not be answerable for any DEBTS to be incurred by my wife ELIZABETH HOLMAN from this date as I have made a separate allowance for her maintenance.
> Dated 30 September 1858. Witness, E.E. Edwards junior.

While by today's societal norms this notice may seem odd, in the 1800s it was not an unusual practice. These "disclaimer notices," or "marital notices" were not uncommon and were a legal mechanism to protect the husband from incurring unreasonable debts while he was away. It does, however, demonstrate the imbalance between husbands and wives at that time, which faded away as women's legal rights improved in the late 19th and early 20th centuries. It is curious though that it would seem he went away in October

1858, only having recently returned. As we shall see later, he would have been away for a very short time.

Further, in writing to John Christoe on 4 December 1859 from Whangarei, New Zealand he said:

Last year I had the offer from Messrs John Taylor & Sons to manage some rich Malachite copper mines on the west coast of Africa near the equator at £600 a year, but refused it on account of the bad climate …

I however had an appointment under the same gentlemen in connection Messrs Bolitho of Penzance and Enthoren of London and went to Malacca and explored some territory for its tin deposits. Before my departure thither I was admitted into their tin smelting establishments to get a smattering knowledge of tin assaying and smelting.

Engaged by Bolitho of London, renowned tin smelters located in Chyandour, Cornwall and H. J. Enthoren & Co. tin smelters of Charlestown, it is likely the assignment would have been for a minimum 12 months. If that was the case, then he would have left England around mid-1857, having returned to Cornwall from South Africa in February. Contacted on his return by John Taylor & Sons — also involved with the Cape of Good Hope Mining Company — he was offered two positions. He accepted the Malacca assignment and would have had plenty of time to visit the tin smelting operations of Bolitho of London, located in Chyandour and H. J. Enthoren & Co. in Charlestown. Both of these locations were within 25 miles of where he lived in River Street, Truro. As both of these organisations were heavily involved in tin smelting, they saw the emergence of Malaysia as a new opportunity for further development and the introduction of Cornish technologies. Furthermore, Cornwall, a dominant player for many decades, was now finding the amount of tin being mined was slowly declining as was its place in the global tin market.

The quality of tin being exported from Malacca had improved significantly from its previous poor reputation. In a period of limited superficial alluvial mining operation, an influx of Chinese from the north was able to increase significantly tin output with new mechanised technologies that allowed deeper mining and more advanced water sluicing operations. Advances in the canning of foods also put huge demands on the use of tin. Combined with declining output from Cornish mines, from the mid-1840s to the mid-1850s tin imports to the UK rose from 4,000 tons to over 21,000 tons. This was helped in 1853 when the British Government lowered import duties on tin and improvements in shipping times from London to Singapore were reduced to just five weeks. Such was the demand that prices of refined tin rose from £105 to £130 per ton from 1852 to 1856. Given all this, is it any wonder that the tin smelters of Cornwall saw Malaysia as a significant frontier that needed to be explored further?

New Zealand, Abbey Farm

1859–62

Only Josiah and Elizabeth know what their motivation was to leave family and friends in Cornwall. One can only ruminate at the many possibilities of their pondering. Was there an agreement between them that he should spend more time with his family? He observed first-hand the terrible conditions miners worked in, the deterioration in health, unforeseen accidents: is this what they wanted for their children? Or, being at the forefront of the mining industry, could he foretell the future? If he could, if he did, then his insight was as close to perfect as it could be, as mining production having peaked in the late 1850s, began a half century of decline.

"The great days were over. Companies folded up. Mines closed down. Hundreds, then thousands, of tin and copper miners found themselves out of work, without hope of employment ... there was no alternative to starvation ... but mass emigration. A third of the population left Cornwall before the end of the century, while back at home their mining towns and villages were left unpeopled, the mines themselves deserted." (Daphne du Maurier. *Vanishing Cornwall*, 1967: 106).

Having returned from South Africa where they enjoyed a milder more temperate climate, the family was attracted to an advertisement in a London newspaper. The colony of New Zealand was offering freehold land to new settlers — if they could pay their own way — or assisted passage for others, especially those with skills. It was suggested that passage to New Zealand would cost around £16. The General Agents to the Provincial Government of Auckland, Messrs Alex. F. Ridgway & Sons of 40 Leicester Square, London, were offering a free gift of a 40 acre farm. Ridgway was one of a number of agents established in Britain, Ireland, Canada and Cape Town during the period of the scheme, 1858-1868, when over 14,000 land orders were issued. By the time the scheme concluded, it is estimated half of Auckland's population were recipients of the program.

---

6   ADVERTISEMENTS.

EMIGRATION TO NEW ZEALAND.

Free Gift of a Forty Acre Farm.

Messrs. ALEX. F. RIDGWAY & SONS

General Agents to the Provincial Government of Auckland,

ARE AUTHORISED TO GIVE THE FOLLOWING NOTICE :—

Every industrious Man or Woman of good character, and not through age, infirmity, or other cause, unlikely to form a useful Colonist, will, on approval, receive a *Free Gift of Forty Acres of Good Land*, in the Province of Auckland, New Zealand, together with *Forty Acres more for each Person above Eighteen Years*, and *Twenty Acres for each Child above Five and under Eighteen Years of Age*, whom he may take with him to the Colony.

The Order for this Land, which will be chosen by the Emigrant himself, will be delivered to all eligible applicants, by Messrs. ALEX. F. RIDGWAY & SONS, General Agents to the Provincial Government of Auckland, and Emigration Agents under the Auckland Waste Lands Act, 1858, at their Offices, 40, Leicester Square, London, on payment of a Fee of 10s. for each Forty Acre Grant, and 5s. for each Twenty Acre Grant. The only condition is, that the party shall have engaged to find his own way to the Colony, and this he can do at an expense of about £16 for passage, on applying to one respectable Ship Broker engaged in the New Zealand trade.

Agricultural Labourers and Servants, Mechanics of various kinds, Domestic Servants (male and female), and Farmers, and others, with Capital (although small Capital), are more especially eligible; but Young Men in the rank of Gentlemen, who have not been brought up to any profession or business, or who are without sufficient Capital to establish themselves in the Province, and to employ labour, are not generally desirable as emigrants, and will, under existing arrangements, not receive these Grants.

Where a considerable number of persons may desire to emigrate in a body, and form a small community in the Province, the Provincial Government will favor such a community, by endeavouring to provide superior Lands and an advantageous location. The Agents will be happy to communicate with any parties desiring to form such a community.

40, *Leicester Square,*
London, 1st August, 1859.

Freehold land was offered on the basis of 40 acres if you were over 18 and 20 acres for those aged five to 18. After much discussion, the family decided they would take up the offer. Josiah travelled to London and visited the agent to complete the application. He was also shown a map and could choose the area that was available as part of the scheme. He chose an area 100 miles north of Auckland, the provincial capital, that seemed to have an easterly aspect over a bay, lot number 10 at Parahaki comprising 233 acres. On the basis of the advertisement, he should have received 80 acres for the two adults and a further 80 acres for the four children over five, totalling 160 acres. But a gift from a Government is never free, there was an application cost of 10 shillings for each 40-acre grant and 5 shillings for each 20-acre grant. It was the aspiration of many, that owning property was essential to attain financial security. Christmas was both a joyful and unhappy time, as they bade farewell to family and friends, Josiah well aware that this would possibly be the last time he would see his parents James and Grace.

On the 11 January 1859, before leaving, Josiah wrote to John Penrose Christoe of Truro, Cornwall:

Mr William Henry Christoe (elder brother) informed me about 3 months since (on my return from Singapore) that you were at Truro some time previous to your departure for Australia when you stated that a Mining Agent would be wanted at the mines which you superintend. Your brother mentioned my name to you then and within 3 months he received information that an Agent would be required shortly after. I wrote to the London Agents about it and they were favourably disposed to me but the following mail brought intelligence that the appointment was to be made on the spot.

I have been known to your brother several years – was once under his tuition getting a smattering of assaying previous to my going to South Africa. I have been abroad six times viz. – to the Philippine Islands, the Brazils, twice to the Canadas and the Native Copper Mines of Lake Superior – to South Africa and lastly to Malacca.

I go out (to NZ) entirely on speculation and failing to get a situation shortly after my arrival in that country I intend to begin as a farmer. Having taken the liberty through Mr. W. H. Christoe's kindness to me to write to you, – my chief object now is to offer my services to your notice and should you see a situation available for me in either the mines under your superintendence or with any other company I should be quite disposed on my arrival at Auckland and receiving intelligence from you of a situation to "board ship" for Sydney. I know no persons at Auckland and neither have I any letters of introduction – hence I go out a stranger among such.

In the letter, he also refers to other offers of employment with John Taylor and Sons, London in late 1858 to manage a silver mine in Central America, but turned it down as the remuneration was not sufficient to change their mind to travel to New Zealand. He suggested Christoe address a reply to the Auckland Post Office.

After Christmas, Josiah contacted the *West Briton* newspaper in Truro and placed an advertisement advertising an auction to be held on 17 January 1859 at his residence in River Street Truro:

Unreserved Sale of New and Modern Household Furniture, Cottage Piano-Forte, Oil Paintings, &c., &c.

MR. JAMES

IS instructed to SELL by AUCTION, on MONDAY the 17th day of January inst., by 11 o'clock in the forenoon, at the residence of Capt. JOSIAH HOLMAN, River-street, opposite the Savings' Bank, *Truro*, the whole of his superior

HOUSEHOLD FURNITURE, VIZ :—

Dining and Drawing-room Furniture, the Furniture of 4 Bed Rooms, Kitchen and Culinary requisites.

This offers a rare opportunity to parties about to furnish, as the greater part have been bought new within the last Eighteen months, and will be unreservedly sold as the Proprietor is about to leave England.

The whole may be viewed the morning of sale.

For further particulars see hand-bills, or apply to Mr. JAMES, Auctioneer.

Dated Truro, 6th January, 1859.

In the meantime, Elizabeth and family continued packing belongings they would take to New Zealand. With all of their possessions either sold at auction or packed in large trunks, they left their home in River Street with their smaller trunks and bags for the final time by coach arriving at Falmouth on 10th February. Elizabeth with the children, Elizabeth Simmons 16, John Henry 12, Emily 9, Josiah Jr. 5 and Charles 2 were all very excited when, on 13 February the

1106 ton, three-mastered *Caduceus* under Captain Cass heaved anchor and left Falmouth harbour bound for Auckland. The *Caduceus* had undergone extensive refurbishment after sustaining damage while serving during the Crimean war. Now a stately roomy vessel, it had been adapted for passenger trade, providing comfort and safety for the long ocean voyages it would undertake.

Light winds prevailed until they crossed the equator on the 12 March heading for the Cape of Good Hope where a gale forced the ship to seek refuge closer to the coast. For the young children, it was an adventure every day. Not so wife Elizabeth, the rocking and rolling of the ocean making her nauseous almost every morning. And eldest daughter Elizabeth seemed to have caught the eye of one of the ship's stewards, George William Robson who was paying her far too much attention to the detriment of the other 300 passengers.

Stopping at Cape Town to replenish freshwater and provisions, a few passengers disembarked, and new passengers boarded for ports East. From Cape Town, a strong southerly pushed the three-masted vessel at a fast clip reaching the west coast of New Zealand on 13 May. Their non-stop, almost 50 days at sea took its toll — two young children died on the way. There were also three births. To the Holman family's delight, it was determined Elizabeth's morning sickness was due to her pregnancy, not the heaving ocean. Passing New Plymouth, they knew their voyage was nearing an end when they sighted the Three Kings Islands and shortly after glimpsed Cape Reinga, then slowly the whole of the tip of the North Island filled the horizon. Travelling down the east coast as they passed Whangarei Heads and Bream Bay, Josiah pointed out the area where he thought the property might be located. But all the children could see were hills and valleys of thick green dense forests. Nearing Auckland harbour, light winds prevented the *Caduceus* from entering the harbour, but an easterly picked up in the afternoon allowing the *Caduceus* to dock at the wharf on the high tide on the 19 May. Pushed hard by the strong southerly, the trip had taken 95 days in all, almost eclipsing the fastest voyage of 92 days. Disembarkation was slow, with over 300 passengers for officials to scrutinise documents carefully and make health checks before coming ashore. It would be another day before the ship was fully unloaded with its cargo.

Buildings were scattered around the harbour and down hillsides that ran, in some instances, quite steeply to the harbour. Along the waterfront on both sides of the docks, a number of store warehouses and commercial buildings, mainly of timber, were being constructed. Behind those, stately homes and hotels dotted the quiet streets. It was not the hive of activity Josiah had expected, with only a few carts and carriages in transit. Once the family had settled into Newell's Auckland Hotel in High Street, he made enquiries on getting to Whangarei some hundred miles north. A small vessel, the cutter *Petrel*, was to leave in two days. The family spent the next two days buying provisions. Josiah also made enquiries on how to get to his property from Whangarei and what he was likely to find there. It seemed the shortest route was to cross the Hatea River and head towards Parahaki hill.

Having settled the family into accommodation, with some provisions and camping equipment, Josiah caught the 20-ton schooner, *Petrel,* on a bitterly cold morning to Whangarei and with map in hand made his way around the bay. Having examined the range and suggested route, he checked with locals and decided instead to follow the Bay of Islands Road through Paranui. After crossing Stony Creek, he headed uphill along the winding road for almost two miles until he came to Lot 10. Being an elevated site, it offered a splendid view over the bay although there were rugged undulating valleys of extensive wilderness between where he stood and the sea. Large boulders and rocks of irregular shapes and sizes were scattered over the property which would make cultivation particularly difficult. Examining the rocks closely he determined they could be cut and fashioned into stone suitable for building. Exploring the property further, he determined much of the property was untamed wilderness with only pockets of level terrain separating dense forests and meandering streams. There were however, enough level areas which would give him a choice where to build a home.

If not for the cold, he would have stayed longer and enjoyed the view, but he left before the fading light closed in on him. Back on the coast, he found accommodation for the night and the next day brought provisions and additional blankets to allow him to stay on the property for longer periods, weather permitting. Having determined a site for their home, he spent the next few weeks selecting suitable stone and began erecting the outer walls. On 30 June, he caught the 20-ton cutter, the *Ant*, back to Auckland. Josiah returned to much excitement in the Holman household as the impending marriage of his eldest daughter Elizabeth Simmons to George Robson was to be held in just a few days. They were married in St Matthew's Church Auckland on 2 July 1859. Elizabeth and George stayed in Auckland, living in Durham Street, George having found a position as a storeman. When the celebrations settled down, Josiah explained to the family the extent of the property's rugged wilderness and what he was proposing in regard to their new home.

On 11 July the family left Auckland for Whangarei on the *Petrel*. With the additional help, work on their house continued at a pace and it was soon habitable. There was an abundance of timber in and around the property used both for building as well as cooking and warmth as required. Josiah's time in Auckland had not been wasted. He had contacted stock agents and arranged for a few cattle to be transported to Whangarei. He also let it be known with farmers locally that he would be interested in purchasing cattle should that opportunity arise. Josiah would later write, ... *my time has been fully occupied, firstly looking after a lot of wild land, 233 acres and making preparations for settling my family thereon as well as partially stocking the run with cattle and bringing a little of it into cultivation.*

Meanwhile, in Byng in central NSW, John Christoe had finally received Josiah's letter of the 11 January 1859. He replied on 10 June indicating the mining operations where he was working had been suspended and he was unable to offer any suitable employment. Josiah wrote back, ... *I regret your mining operations at Carangara have been suspended – I hope only temporarily.*

Perhaps it was experience or just good luck that wife Elizabeth's pregnancy was proceeding with very little discomfort, while the rest of the family progressed with the home building on Abbey Farm and in one sunny area, developed a garden for growing vegetables. On 15 August Josiah and son

John caught the *Petrel* to Auckland to visit Elizabeth and George as well as buy more provisions, building materials and source fruit trees. When they returned, they were greeted enthusiastically by very excited, children running gleefully and crying out "we've found some caves." While searching for wood in a nearby valley, they had seen an opening in the rock formation and like all little children embarked on a journey of discovery. The entrance was quite small, they had to crawl a little way to avoid the hanging foliage, then clamber over rocks impeding their progress. To their amazement and delight, they found it opened up into a large cave with a stream running through it. Teeming with glow worms, the light cast eerie shadows across the shimmering damp walls, so they decided not to proceed any further but would come back with a lantern later. Returning with their father, they descended further, the light allowing them to proceed, but with caution. Continuing along the damp floor, they crawled through an opening which revealed a huge cave almost 40 ft high. Stalactites hung from the ceiling and when Josiah tapped them, they rang like a church bell, much to the amusement of the children. It wasn't so dark in this huge cave, tiny glow worms glittering much of the ceiling. Josiah knew well the adventurous spirit of the children would drag them here every day, if they were allowed. As well, it was not so safe and he was unsure just how far the cave network extended, he would need to set boundaries for the children.

The weather was warming in late October when Elizabeth announced she was feeling those familiar twinges of pain she had experienced many times before. The family gathered up what was necessary, blankets, sheets, plenty of firewood to keep the water boiling and on 2 November 1859 daughter Annie was born. Josiah and of course the children were beyond excitement, especially as all had gone so splendidly well with the birth.

A few days later when all had calmed down, Josiah wrote to his parents, John and Grace, and rode the four mile journey downhill to the Post Office at Whangarei. It was only by chance while waiting outside that he noticed the list of unclaimed letters on the window including one redirected from Auckland addressed to a Joseph Holman. He enquired inside asking if the letter was perhaps for him. The postmaster found the letter and on examination noted it was from a John Christoe. Regrettably, the letter did not bring the news Josiah had been eagerly anticipating. He replied on 4 December 1859 explaining that he would be interested ... *if anything worthwhile turns up in NSW ... it is however pleasing to find by your letter that mining matters had taken a lively turn. The lode you refer to in the Canobolas Mts is very rich and unusually wide / a chain / I have seen deposits of copper ore somewhat wider but ill-defined extending not over 3 to 6 chains in length where all traces of ore disappear. Having housed and settled my family in Auckland, I am open to take a situation such as you state – my services probably would command in Sth Australia viz. £500 a year or thereabouts. Such remuneration would induce me to leave this place for 6 months to 2 or 3 years or permanently if the country suited me.*

Work continued over summer on the "wild acres", clearing any level areas, removing stone to the house to build dry walls and develop gardens for growing vegetables. The children were home schooled as best as possible, reading and developing their writing skills.

In early 1861, Josiah was contacted by the Otea Copper Mining Company (OCMC) (Limited), most likely through Charles Bischoff, the solicitor acting on behalf of OCMC. This was the same Charles Bischoff he had met in Montreal in 1853. The Great Barrier Land Harbour and Mining Company (Ltd) (GBLHMC) owned 26,000 acres, about a third of the Great Barrier Island North East of Auckland. A mine on the north portion of the island had a somewhat chequered past, Abercrombies and Messrs Whitaker and Heale owned the mine up until 1857 and had shipped out around 730 tons of copper ore. On the recommendation of John Taylor & Sons, Captain James Ninnis and William Rowe were appointed to manage the mine. They leased it for a year before being approached by the GBLHMC in 1858. After three years' operation, it had become clear additional capital was required to carry out further deeper exploration. Under the direction of Heale, operations had been suspended and confined to less costly surface work. GBLHMC now wished to raise capital by issuing 10,000 shares of £5 each with a deposit of 10s. per share at the time of application and 20s. per share upon allotment. The prospectus of the new company, Otea Copper Mining Company (Ltd). further stated about 800 tons of ore had been mined, and sold at Swansea for around £13,000, with an average price of £16 5s per ton. At a meeting of shareholders on Tuesday 5 March 1861 at Bishop Gate Street London, it was proposed to …

> transfer the mine to another company on terms which your directors are of opinion ought to be carried out … his company had no available funds for properly developing the mine.

By these means, GBLHMC would receive a considerable sum to improve its extensive property without making a call on the shareholders and would still retain a large interest in the mine in paid-up shares and also receive a royalty on ores raised. Mr Heale further recommended that "… a report from one or more competent and independent mining captains" be obtained.

Josiah's name was put forward and references requested. Among other testimonials in his favour, was one from Humphry Willyams, Banker of Truro, a partner in the well-known copper smelting firm of Messrs Sims, Willyams and Co, attesting to Josiah's many years as Mr Willyams "confidential mine agent":

> It gives me great pleasure to reply satisfactorily to your inquiry about Capt. Holman. I have known him for a great many years and have employed him on many occasions and in all parts of the world. He is an extremely intelligent, judicious and trustworthy man – sober and honest to the fullest extent, and I consider him to be fully competent to be entrusted with the management of any mining undertaking.
> H. Willyams, Carnanton, 5th October, 1861.

Willyams wrote again on 23 October, 1861:

> I sincerely wish for your success in your proposed undertaking, and I take the present opportunity of confirming my previously expressed opinion

of the judgement and ability of Capt. Holman of which I have had many years' experience.

And on 6 November, 1861:

You are at perfect liberty to publish my letter respecting Capt. Holman. My opinion of him exceeds what I have expressed on paper.

An abridged prospectus was later issued proposing that capital would be raised by the sale of 25000 shares, of £2 each, 5s per share to be paid on application and a further 5s per share on allotment. Under the provisions of the Parliamentary Act, each shareholder's liability would extend solely of their subscribed amount. Josiah was contacted and asked if he would be available to examine the mine and provide an expert report. He was not about to turn down paying work, the £500 he had brought with him was dwindling quickly:

The expenses incurred in bringing my family out and settling them etc. will I fear soon necessitate me to look out for a situation till such time as my land makes me a sufficient remuneration to enable me to remain on it comfortably. I do not expect much return from it for 2 or 3 years and in the meantime, some hundreds of pounds ought to be spent on it more than I can at present command.

The Great Barrier Island was not far from Whangarei, lying further to the east, halfway between where he lived and Auckland. In March, 1861 he boarded a small cutter and spent a week examining the mine and works that had been undertaken. He inspected a number of the stopes, adits and a winze which had been sunk from the surface to meet the lode. He noted that at the 12 fathom level, the quality of the ore continued, not dissipating, which was often the case; a good sign. A tramway had been constructed to carry the ore to a crusher driven by a small engine. Adjacent were a number of spacious workshops to dress the ore which was then conveyed to the dock. About 60 tons of ore lay at grass ready for shipment.

A random collection of cottages and other buildings were scattered about at the head of the bay forming a small village. In Captain Holman's report of 21 March 1861, he stated:

Over 3000 tons of ore, fully 15 per cent for copper, available above the adit level, most of which I think could be stoped for 12 shillings per fathom, the whole size of the vein. By adopting such a system of stoping on a much larger scale than hitherto, both above and below the adit, by force of about 20 miners and 40 labourers, it would, I confidently believe return sufficient ore to meet the expenditure of the mine. A force of six men could be placed partly to drive the adit level South, a very desirable object, and the remainder to drive the 12 and 22 fathom levels. The principal test approving the mine in depth should mainly depend on the stoping. With the above force for mining, and the corresponding staff on the ore floors, I consider about 50 tons of ore, of 15 per cent, could be raised monthly, or two tons per working day, at a cost on the local establishment from £500-£550 per month, exclusive of rates, insurance, sales of ore et cetera.

In conclusion, I may state that the mine may yet be said to be of a spectacular character on the whole. Still, there is a proof shown by the deeper explorations that the ores are not merely superficial, but that it is a vein that will evidentially continue in depth. Whatever course of action is determined on by the company, the mine fully warrants an extended trial.

In mid-May, Josiah again visited the island and in his report of 28 May wrote, *If only a permanent increase in the yield of ores takes place throughout the vein – such as seen in the 12 fathoms level, where the quality of the ores is quite equal to the general shipments – the future value of the mine would be very great.*

Asked to visit the island again in November, he wrote on the 27th increasing his estimate of the possible yield of the ore to 4000 tons, if the powerful crusher he recommended were erected:

The materials required will be few in number. Steel for borers, with powder and fuze, include the chief items for quarrying. On the dressing floors steel hammers and sledges, with riddling and jigging sieves; for the crusher the usual wearing parts; while, for the engine, wood is abundant and easily procured.

In the prospectus was added, "The value to be attached to Capt. Holman's statements and opinions can be inferred from the very high testimonial in his favour by Mr Humphry Willyams, the well-known copper smelter, given at the end of this prospctus. It will be seen that Mr Willyams expresses himself in strong terms, founded on his knowledge of Capt. Holman for many years, and his employment of him on many occasions in all parts of the world."

And … "As soon as the necessary capital is subscribed for, the requisite machinery will be ordered and sent out. It is also proposed to appoint Captain Holman the Managing Agent of the operations, which he has offered to undertake at a reasonable salary, and as he is already in the colony the expense of sending out an agent will be saved, and no time be lost in carrying out the objects of the company."

In April 1861, Josiah added his name to the list of landholders requesting his property, Parahaki freehold lot 10 be registered. In June, his daughter Elizabeth and husband George announced her pregnancy, the baby due at the end of the year.

There was much excitement in the colony in September 1861 with the anticipated pending arrival of Sir George Grey returning as Governor after a seven-year appointment as High Commissioner in South Africa. He was well known to Josiah as it is very likely they had met when he was leading the mineral expedition in Namaqualand for the Cape of Good Hope Mining Company. In October, Josiah called on Government House in Howick, 10 miles east of Auckland harbour and presented his testimonials in the hope of being considered for the position of Inspector of Mines. On 29 October, Samuel Sneddon wrote to William Fox the Premier and Colonial Secretary in Auckland:

My dear Sir

A person of the name of Josiah Holman, now residing at Wangari, lately called here offering himself as inspector of mines to report on gold fields and such like.

His Excellency Sir George Grey desires me to name him to you, and I conceive this is the best way of doing it …

Letter reporting on Captain Holman's visit to Governor Grey, Archives NZ

In late 1861, Josiah gathered a number of specimens of stone, lignite, lava, and quartz to be added to Auckland's contribution for the International Exhibition to be held in London in May 1862. On 8 January, 1862, a son Charles was born to George and Elizabeth at their address, Chapel Street Auckland. This was Josiah and Elizabeth's first grandchild, more than 30 would follow over the coming decades.

Meanwhile, in early 1861 in Australia, the Scottish Australian Mining Company (SAMC) had taken over the Oakey Creek mine at Cadia outside

Orange and renamed it the Cadiangullong mine. John Christoe was engaged as mine manager to develop and manage assaying for new smelters yet to be built. On the Sydney docks, a 25-inch engine and crushing machine ordered by SAMC for the Good Hope Mine near Yass, about 170 miles south-west of Sydney had been gathering dust since its arrival on 9 May 1860. Ordered from Harvey & Co. of Hayle by SAMC, its delivery to Yass had been deferred awaiting proof of the lode's quality. With the improved output and quality of the Cadiangullong mine, it was decided in late 1861 to transport the engine to Cadia; but they would need someone with considerable experience to manage the design and building of the furnace, installation of the boiler and ancillary equipment and oversee the operation of the massive machinery.

Christoe suggested Captain Holman, living in NZ and keen to secure work in Australia, be appointed; he was after all, much closer than bringing someone out from England. Robert Morehead, manager of SAMC was aware of Captain Holman, having followed New Zealand's various mining projects after the joint Kawau Company mine failure a few years earlier. He agreed, and on 7 January 1862, Christoe wrote to Josiah, offering him the mine manager position at the Cadiangullong mine, salary £500. Upon receipt of the letter, Josiah immediately wrote accepting the offer and contacted a shipping agent in Auckland to book the next berth to Sydney. Elizabeth and the children would stay, Josiah wishing to see what accommodation there would be for the family, what amenities were in place and also determine the likely long-term viability of the mine and therefore his tenure. The family was quite used to his comings-and-goings, but this may well be a little longer than usual. In early March, Josiah caught a small cutter to Auckland and on 8 March boarded the 286-ton *Airedale* to Nelson and then to Sydney.

In mid-1864 Elizabeth received a letter from Josiah describing the ongoing development of the village at Cadia and the house he was developing as the mine manager. Progress was continuing rapidly with the development of the mine and the blast furnaces were all now operational. The village of Cadia was growing with the influx of over 400 miners and other workers and families; a variety of shops and a new tavern were almost completed and a National School was being proposed. A post office had been established and a mail service ran once a week from Carcoar to the growing township of Orange, which also boasted a telegraph service. It was agreed Elizabeth and family would return later that year. In mid-December, Elizabeth and family with all their possessions caught a small cutter from Whangarei to Auckland where they stayed over Christmas with daughter Elizabeth and George Robson and grandson Charles. On 3 January, Elizabeth, Emily 15, Josiah Jr 10, Charles 7 and Annie 4 boarded the newly commissioned 600-ton *Egmont* under Captain Tredwen and left at 12.30 PM for the port of Sydney arriving on the 11 January 1865.

Passenger list of Egmont from Auckland to Sydney arriving 11th January 1865: Mrs Holman, Miss (Emily), Miss A (Annie), Master J (Josiah), Master C (Charles).

John Henry stayed on at Abbey farm and having learnt much from his father, continued mining trying his luck when gold was discovered in the Thames region in 1867. He married Mary Josephine Keefe at St Patrick's Cathedral Auckland on 30 December 1870. He passed away on the 18 January 1897. Abbey Farm property was rented to Mr Humphreys when John Henry was away, and on the 9 August 1866 the house and its entire contents were burnt to the ground. In December, a number of the articles from Abbey Farm were found in a home of a Māori who was being pursued for burglary of another home. Police concluded the Abbey Farm house was burnt down deliberately to hide evidence of the theft.

Josiah continued with petitions to the Auckland Provisional Council *praying* for a land order. On 23 November 1866 Mr Rows presented a petition and on 8 December 1868 Mr McCready again presented a petition on Josiah's behalf and on 17 February 1869, Major Cooper moved:

"That a respectful address be presented; to the Superintendent, inviting his Honour to give effect to the interim reports of the Petitions Committee, relative to the petitions of ... Josiah Holman ... and to deal with cases in accordance with clause 30 of the Wastelands Act, 1867".

The honourable member also moved that the reports be read, the motion being seconded by Mr J Graham and was agreed. On 18 November 1870, Mr Swanson moved ...

That the petition of Josiah Holman which was presented to this council in session 24 and favourably reported on by the Petitions and Private Grievance Committee in its interim report number nine brought up on 17 February 1869 be referred to the Petitions and Private Grievance Committee.

It was again agreed.

On Tuesday, 4 June 1867, at Austinfriars, London the Otea Copper Mining Company held an extraordinary general meeting of shareholders to report that ...

The quantity of all ore received was 24 tons which is now at Swansea, being prepared for early sale. Another parcel is on the way and a third small shipment would follow. The quality is believed to be considerably higher than was expected.

Mr Allom, the general agent of the company, had employed Captain W. Rowe to make an examination and recently forwarded his report to the directors. The report explained that Captain Higgins was unable to deal with the difficulties that had arisen in dressing the ore and this attributed to the delay in making large returns. Captain Rowe stated …

> The stopes both North and South are looking well, I have never seen them looking better and although I have always thought both Captain Holman and Higgins have overestimated the profits that might be made, yet I always have believed and do now most firmly believe that the Otea copper mine is a most valuable property and will yield adequate returns for the outlay made.

On the 30 June 1867, an extraordinary general meeting of shareholders was held at Austinfriars, London. A resolution was proposed and agreed to authorise the directors to borrow a sum not exceeding £10,000 on the security of the whole of the property of the company. This would be raised by inviting the subscription of £10,000 by the shareholders, that sum being considered by the directors necessary to pay the liabilities and carry out additional works. At an extraordinary general meeting of shareholders held on 30 October, the directors, after full consideration, determined that the best course was to wind up the company. After some discussion, it was resolved that the Otea Copper Mining Company be wound up, and that a special meeting be held on November 28 to confirm that resolution.

But worse was yet to come, when, in early February 1868 a fierce storm struck Great Barrier Island. There was considerable damage to the wharf and mining operation to the extent that some machinery was washed into the sea. On 25 February, in the *Daily Southern Cross* newspaper, a notice appeared advertising an auction of the entire machinery and plant of the Otea Copper Mining Company (Limited). On 20 March, an auction of all the plant and machinery of the Otea Mining Company was held at the stores of S. Cochrane and Son, Fort-Street, Auckland. The auction attracted considerable interest and a large attendance resulted in brisk bidding. The engines and a great portion of the heavy plant was sold in one lot for £800 to Captain Ninnis. Bar iron, steel, tram wagons, anvils, bellows, brass wire grating, hammers, sledges and a large variety of tools and materials all brought fair prices. In January 1869, Frederick Wollaston Hutton, geologist, presented a geological report on Great Barrier Island. In that report, he estimated the total amount of copper ore that had been mined up to 1869 to be 2323 tons.

To May 1857, Abercrombies, Whitaker and Hale – 783 tons.

To July 1858, Ninnis and Kowe – 745 tons

To July 1861, Great Barrier Company – 574 tons

To December 1867, Otea Copper-Mining Company – 271 tons

PART FOUR
CADIA
1862–1893

The Beginning

It may seem a little incongruous to begin a chapter on Cadia, just outside Orange in New South Wales, with the story of an Aberdeen company in Scotland. However, the fortunes of this company, and the working life of Captain Josiah Holman over two decades are inherently interwoven.

In 1840 in Aberdeen, The Scottish Australian Investment Company (SAIC) was formed to raise capital of £100,000. The company's objective was to invest funds in the burgeoning colony by "making loans on heritable, and other securities, in discounting bills, and in making small and judicious purchases of land."

Banks in the British Isles were loaning money at around 6 percent whereas in the colonies 10 percent was common, illustrating the sound commercial opportunities available.

A prospectus was issued in mid-1841 and 100,000 shares of £1 each offered to the public. Directors, mainly wealthy businessmen of Scotland, were appointed. Mr Robert Archibald Alison Morehead was engaged as manager and Mr Mathew Young as sub-manager. Robert Morehead had a strong finance and accounting background having worked as a book-keeper and accountant in Glasgow. The two managers and their families left Plymouth on the barque *Abberton* of 481 tons under Captain Catt on 6 April 1841 arriving in Sydney on 21 July. They quickly went about establishing offices at 5 Fort Street Sydney and by early August were advertising that they were open for business.

> **THE SCOTTISH AUSTRALIAN INVESTMENT COMPANY.**
> **CAPITAL—**
> **One Hundred Thousand Pounds,**
> **with power to add to it.**
>
> The Manager of the above Company will be ready to receive proposals for Loans, on Monday, the 9th instant.
> 5, Fort-street, Sydney, August 6, 1841.

Sydney Morning Herald 7 August 1841

In late 1841 a director position became available at the Sydney Banking Company when Mr R G Dunlop retired. At an annual meeting of shareholders on 12 January 1842 Mr R A Morehead was elected director. The Sydney Banking Company held deposits totalling £37,000. It also had notes in circulation to the value of around £17,000 and stock estimated at around £150,000. The bank had increased its bullion to nearly £26,000 in coin and Bills Receivable amounted to nearly £200,000. For the half year to December 1841 a profit of almost £9000 was declared. While Robert Morehead's appointment seemed a fortuitous move at the time, within a year he had discovered irregularities in the reporting by the accountant. Further investigation led to arrests in mid-1843 and the following year the bank was forced to close its doors. Despite the bank's closure, this adversity substantially enhanced Robert Morehead's credibility within the business community.

For venture capitalists, the depression of the 1840s provided rewarding opportunities for the discerning investor and by the mid-1840s, SAIC had built an impressive property portfolio in and around Sydney with a solid book of mortgages from £80 to £3000. They had also moved from Fort Street to larger, more impressive offices in "respectable" O'Connell Street. To further establish his business credentials and respectability, Morehead immersed himself into the select echelons of society of Sydney.

In 1844, there was much excitement in the colony when a rich copper vein was discovered at Burra Burra in South Australia. Miners from Cornwall and Wales flocked to the area and by 1846 a small township of more than 200 had risen from the almost barren desolate landscape. The Burra Burra mining area was a little over 100 miles north of Port Adelaide, six days by bullock dray. Surrounded by low elevated hills, and split through the centre by Burra Creek, immense piles of ore were soon scattered throughout the 50 or so acres. Next to them, shafts were being blasted or dug deep, many with five or six horse whims placed over them, some working day and night. During the day, whims raised ores from the sunken shafts, some of which were over 320 fathoms deep, or raised water in barrels to wash the extracted ore. The immense richness of the ore — up to 80 percent it was reported — excited miners seeking their fortune as well as wealthy investors looking for quick returns. The largest and most productive by far was the copper mining company South Australia Mining Association (SAMA). Their profits soared and by 1849, their 2464 shares issued at £5 were worth a staggering £152 each. Declared dividends, over six times the issue price, was not uncommon.

Burra Burra copper mine circa 2022 – Geoffrey Quayle

In April 1846, Richard Morehead visited South Australia and paid £4950 for 347 acres bordering the prized Burra Burra mining lease. While Morehead's remit was wide — maximise returns on capital, mostly charging 15 percent interest on mortgages — this was the company's first foray into mining in

Australia. A mining company was registered, Bon Accord Mining Company, generally referred to cheekily as "Sydney's Mining Company". Morehead wasted no time in proceeding to maximise the investment for the joint owners, SAIC and the Aberdeen-based North British Australasian Loan and Investment Company and two weeks later advertised for miners to begin exploration. But finding miners with experience would prove difficult, even the huge influx from overseas could not meet demand. However, at the end of 1846 the company reported it had sold ore exceeding £45,000 and would likely deliver a profit of over £64,000 in the first year.

The Kapunda / Burra Burra area continued to thrive. By 1848, a large smelting house built of stone with storage for charcoal was taking shape along Burra Burra Creek. Huts for some 500 miners dotted the hillside and a brewery making excellent ale stood at the northern end of the creek. Just over the hill, the small township of Kooringa was growing rapidly as people poured in attracted by employment and opportunities of a mineral find. Kooringa soon supported over 5000 people and grew to become the largest inland settlement in Australia. It was however, a company town. The South Australia Mining Association's huge profits provided capital for tenanted homes, stores for all manner of supplies as well as a bakery, butchery, Inn and a Wesleyan Chapel was built in late 1847. Three dairies operated on the outskirts. But there was no school. With adult workers mostly engaged in mining, much of the menial work was carried out by youths aged from 10 to 15 years. Missing too was a medical professional. In 1848 Dr Matthew Henry Smythe Blood, born in Limerick, County Clare, Ireland on 6 December 1806, arrived in South Australia with his wife and family. They travelled 100 miles to Kapunda, halfway between Adelaide and Burra Burra, where Dr Blood was appointed the Mines Doctor.

Meanwhile, Captain Spargo whom Morehead had engaged as mine manager, after some initial success, was struggling to trace the continuation of the reef of the very successful Burra Burra mine adjacent to the Bon Accord property. The rich vein he had hoped would continue through to the Bon Accord lease was not to be found and in late 1847, after a visit by Morehead, the mine was closed. As time progressed, it was becoming clear this first foray into mining in Australia was not going to be the success Morehead had hoped. In early 1849, the district was visited by the Surveyor General and the township of Redruth was established. Also, an area which included the Bon Accord acreage was selected to be developed as a township which would be named Aberdeen in deference to the Scottish address of SAIC. Immediately, 144 lots were drawn up for sale in hope of recouping some of the considerable investment. Richard Morehead attended an auction held in Aberdeen on Saturday 22 September 1849 where lot sales generated £500. Later, in March 1851, Morehead, his wife Helen, two children and a servant travelled by ship to Adelaide and then to Aberdeen. They spent just over three months reviewing the township's layout, allotments and progress as well as inspecting the mine and discussing options with mine managers before returning to Sydney.

In late 1851, SAIC acquired 400 acres near the town of Yass about 170 miles south-west from Sydney which included the Good Hope Copper mine. They had engaged Captain John Dalley, of St. Austell, Cornwall who had worked in Brazil and at Burra Burra to examine the mine and in his report dated February 1852 he noted that: "I do not wish to raise hopes to much in respect to this property, but surely I look upon it as one of great promise."

In September 1852, John Penrose Christoe, who had arrived in South Australia some years earlier from Wales, married Dr Blood's eldest daughter, Dorothea Juliana, born in 1836 in Springfield, Ireland. The following year Christoe and Blood put their names forward to be elected as members of the Lower Chamber for the Counties of Light and Eyre, South Australia. In 1854, possibly at the recommendation of his father-in-law, John Christoe and wife Dorothea returned to Wales so he could gain greater technical experience as a mining assayer. John Christoe's father, Bill, was a highly regarded assayer, as was his elder brother William, an assay master at the huge Swansea Smelter. Swansea smelters processed much of the world's ore, their expertise and "secrets" so dominant.

While the ventures into mineral exploration for SAIC were not delivering the results as hoped, mortgages and investments in property around Sydney, including the strategically located Bon Accord stores on the harbour foreshores at Circular Quay, were delivering positive returns. Almost every day they advertised in the Sydney newspapers, properties for lease or sale as well as tenders to develop land and properties. Business was booming in the new colony. Dividends of around 7 percent were being consistently paid to shareholders and at the half yearly meeting in December 1853, a profit of almost £17,000 was declared for the six-month period. It was agreed a dividend of 15 percent would be paid. So pleased were the directors of SAIC, they wrote "prospects of the company were never in a more prosperous and favourable condition than at the present time."

Richard Morehead continued to promote himself wherever possible, being appointed joint auditor for the City of Sydney, member of the Board of National Education, treasurer of Philosophical Society and Australian Library of which the president was James Mitchell. But Morehead also promoted his vested interests and was a vocal advocate for the development of infrastructure including roads and railways in New South Wales.

By 1857, it was becoming evident that SAIC investments in mining ventures were either exceedingly unlucky or not being well managed. North British Australasian Investment Company, their joint venture partner operating the Kawau mines 30 miles north-east of Auckland, reported a total loss of £78,473 after it had been abandoned. The lack of success in their mining ventures in South Australia, Bon Accord, the Island of Kawau in New Zealand and the Good Hope mine did not completely deter the directors of SAIC. After a lengthy discussion they proposed a change of direction and recommended the following:

"To preserve the character of this Company as an Investment Company, keeping it free from mining adventure, through the medium of separate companies and secondly, to secure to its shareholders the greatest amount of benefit from their mineral investments that may be practicable, without their necessarily engaging in any actual mining operations."

However, there were still those who felt their investment in SAIC were being put at risk by the expensive unpredictable ventures into mining. In early 1859, it was recommended that a company combining the Scottish Australian Investment Company's mining interests and the North British Australasian

Investment Company, which jointly owned the Bon Accord and Good Hope mining properties, would be formed and listed. A prospectus for the Scottish Australian Mining Company (SAMC) was issued in London to raise £75,000 in shares of £1 each. The consensus amongst various mining captains and other "experts" was that if the rich Burra Burra copper seam was to be found, heavier steam power would be required to excavate further than the 20 fathoms achieved thus far, especially given the Burra Burra mine was now down to 50 fathoms. A prospectus was issued stating capital raised was for "… thoroughly exploring and developing a mineral property".

SAIC was to receive £12,000 in cash and 8000 shares of the new mining company plus a proportionate royalty share of all ore raised.

Morehead could see the huge potential this fledgling colony offered and the board agreed. Words across the ocean are easy, but action is an entirely different matter. After numerous attempts to convince the directors of SAIC to raise capital for further expansion in this booming settlement, frustration got the better of him and he withdrew £17,000 for the purchase of various properties and carefully selected investments. He had been expressly ordered by the board that any withdrawals were not to exceed £12,000. At a meeting on Friday 29 January 1858 at the London Tavern, Bishopsgate, London, great dissatisfaction was expressed by the board. Not only had Morehead flagrantly disregarded the boards directive, but his withdrawals had depleted the account to the extent that there were insufficient funds to pay the 8 percent dividend the board had recommended. An embarrassed Chairman William Dixon added critically: -

"It would have been better had he drawn for a lesser amount, because he had certainly crippled their resources, and had put the London board to great inconvenience."

Shareholders too were incensed, and a resolution was moved to pass a vote of censure on Morehead for his wilful disobedience of the orders of the Directors. After some argument, it was eventually withdrawn. To placate disenchanted shareholders, the board agreed to raise the dividend to 10 percent; however, the tone of dissatisfaction had been established.

It was perhaps for this reason Morehead was recalled, and with wife Helen, children and servant, Morehead left on 10 June 1858 having boarded the *SS Australasian* for Southampton via Melbourne. It is unlikely that Morehead would have objected, although he might have been fearful about his position with the company. He had not seen family and friends for 18 years and there was another benefit. It would provide a great opportunity to convince the "doubters" of the boundless potential available in this new colony. Also, on the 26 January 1859 at the Albion Tavern, 153 Aldersgate Street, London, there was to be an Australian Anniversary Dinner celebrating the 71st anniversary of the settlement which would be attended by over 150 dignitaries. The importance of such an occasion was not lost on the canny Scot as it would present another compelling chance to promote the great commercial and investment opportunities in Australia. The Albion was a fine establishment of four stories renown for holding extravagant dinners of the highest quality. But before the dinner, there were many meetings scheduled on his arrival in London, and after attending the SAIC Annual General Meeting in January 1859, the Chairman W H Dixon wrote:

"The visit of the manager to England, after an absence of nearly 18 years in their service, has been productive and of much advantage to the company. The board have been materially assisted by him in effecting the disposal of the mineral properties which have been sold since his arrival, and every opportunity has been and will continue to be taken during the remainder of his stay in this country to consider and discuss with him personally all such matters as may be thought capable of being made conductive in the future prosperity of the company."

While these words were certainly appreciated, Robert Morehead still needed to convince the directors great profits could be realised with further investment. Having achieved the implied backing of the board, he was not about to lose this opportunity, and delivered an impassioned plea for even more additional funds, leaving no doubt about the significance of his request:

> I am anxious to place clearly and prominently before the proprietors that the measure (additional funds) is in strict furtherance of a principal or

object which I have aimed at from the very early period of my connection with the company. If, therefore, my past management has gained the confidence of my constituents, I would beg them to keep in mind that the measure to which I am now advertising rests on the same judgement as that which has been acquired that confidence. It would be matter of regret if, after showing by our successes the soundness of this conception, we are obliged, through want of adequate means, to lag behind the communities in which we have established ourselves.

He further added: "Before bringing to a close the remarks I have now to make, I would express the satisfaction affords me to believe that my visit to this country, now drawing very near to a close, has not been unattended with benefit to the company, and in conjunction with this consideration I would observe that I am confident the frequent and full discussions we have taken place between the board and myself, with reference to many questions relating to the business of the company, will materially facilitate future discharge of our respective duties. Lastly, I have to acknowledge a great deal of kindness and consideration received by me from the members of the company."

His words were well received and finally the board agreed and at the meeting proposed a resolution authorising the issuing of 200,000 preference shares which would raise a further £200,000. It was a triumph for Morehead, not a wasted trip at all. In convincing the shareholders, the chairman further added: "… it was the genius of the company to acquire at a low rate and dispose of it at a high rate."

Another shareholder remarked:

Mr Morehead had been 18 years in their service and had never made a bad investment, which was a sufficient guarantee that they could trust the new capital in his hands.

The directors, in considering the services of Mr Morehead during the long period of 18 years, and the expenses sustained by him in visiting this country, were unanimously of the opinion that the sum of £1000 should be presented to him. The shareholders and directors of the SAMC were equally impressed by Robert Morehead's performance in managing the company's interests. One of the directors remarked:

> They had great reason to congratulate themselves on the excellent judgement which Mr. Morehead had displayed in the management of their affairs in the colony, and he was sure he was speaking of the feelings of all present in offering an expression of their acknowledgement to Mr. Morehead for his past.

The Chairman added, "He had great satisfaction in putting such a resolution to the meeting as he could bear testimony that Mr. Morehead, the manager of the company, who is at present on a visit to London, and was at the meeting, had used his most strenuous efforts to further the interests of the company."

At the SAIC half yearly general meeting of proprietors held at the London Tavern on July 29 1859, the chairman, W. H. Dixon declared a 10 percent dividend and again heaped praise on Morehead's character. His " … advice and unrelenting attention to their interests since he had been in this country have been most valuable to the company. Mr. Morehead had shown great discretion in purchasing property at a nominal price and afterwards realising great profit through his subsequent development and improvement."

With a firm belief that mining on a larger scale was the key to success, Morehead investigated what machinery could be used to achieve this. He had already convinced the conservative directors that the only way to realise great profits was to invest in larger scale operations with machinery. With the board's approval, he arranged for SAMC to commission the design of a 25-inch rotative beam engine, to be built by J Thomas & Company at the Charlestown Ironworks in St Austell, Cornwall at a cost of £1225. It was to be identical to the 25-inch engine then operating at the South Crinnis Mine in Cornwall. The engine was inspected by James Sims, a Cornish mining engineer who specialised in compound engines, on behalf of SAMC before its despatch to

Sydney in late 1859. Sims was well known in Australia having designed two engines for use in South Australia. One in 1850, a 48-inch Bull engine and the second in 1851, a 36-inch Bull engine, both made by Harvey & Co. of Hayle. His report on the inspection of the 25-inch Thomas engine and the crushing machine was most favourable; a copy of his handwritten letter from Redruth still exists, dated 22 September 1859.

While in London, positive preliminary trial exploration reports continued to filter through from the Good Hope mine near Yass. Captain Dalley had been carrying out small exploration trials of the Good Hope mine since 1852. In early 1858, after completing a deeper exploration trial, he reported that "the lode is strong and large", and that an assay yield displayed a result of 21 to 36 percent copper, one piece almost 75 pure copper. In May, tenders were called to erect miners' cottages and by August, there were 10 men engaged opening up a shaft to carry out further experimental trials. It was determined that the 25-inch Thomas engine being built in Charlestown would be transported to the Good Hope mine. On the South Australian Bon Accord mining site, the building of a manager's residence, mining office, blacksmith forge and carpenters' shop were nearing completion. On the 7 September 1859, the Duncan Dunbar left Plymouth with Mr and Mrs Morehead, family and governess on board, arriving after 12 weeks at sea on the 27 November 1859. Returning to Sydney, Robert Morehead, now with the backing of the board, wasted no time in exploring new opportunities.

A little earlier, in September 1858, J.P. Christoe, now living in Wales, was asked to take on an appointment as assayer for the Carangara Copper Mining Company in NSW, near the village of Byng, just east of Orange. The Carangara Copper Mining Company was registered in July 1854 and work began shortly thereafter. Within a year there were almost 50 people working at the mine carving out shafts and with further land purchases, the company now owned nearly 2000 acres around the mine site. Several hundred tons of ore had already been extracted and a small village had begun to emerge with stone and slab homes, a school, stores and a chapel. Christoe assembled a team of assayers, miners and smelting equipment and left England in early 1859. When he arrived, there was much discussion about a new mine to be established by the Carangara Copper Mining Company nearby. Under Christoe's direction, progress was being made with the building of a smelting works at Browns Creek near Guyong, just south of Byng, to smelter ore from the Carangara mine and the nearby Ophir Copper Mining Company and Canobolas mine to the west owned by Mr (later Sir) Saul Samuel and other directors.

Josiah's letter of 11 January 1859 was possibly sent to John Christoe's address in Wales and then readdressed to NSW, as Christoe did not reply until 10 June. When Christoe did reply, he indicating that mining had been suspended and unfortunately there were no work opportunities available. In fact, shortly after in November, an advertisement appeared in the Sydney papers offering for sale a steam-engine, in good working order, with pumping gear attached and about 40 fathoms of lifts at present erected at the Carangara Copper Company's mines, near Guyong. Plagued by exorbitant cartage rates and the delay in building a railway connection that would significantly reduce costs, mining companies west of the Great Divide struggled. Transporting a ton of ore to a smelter in Sydney, at say 25 percent yield of pure metal, significantly increased the real cost of mining. Cartage rates by bullock teams ranged from £4 10s. to £5 per ton, regardless of whether it was ingot or ore. Carting ore added £18 to £20 to each ton of pure metal produced, almost a quarter of its sale price. Smelting ore locally and transporting the ingots was a huge saving for mining companies. Any closing of a mine would also impact the viability of a smelter, vital in reducing mining costs.

On 13 January 1860, the first annual general meeting of the Scottish Australian Mining Company was held at the London Tavern, Bishopsgate Street, chaired by Sir Edwin Pearson. It was agreed at the meeting to *"confine their operations at present to the Good Hope Mine."* The 25-inch rotative beam engine and its associated equipment arrived in Sydney on 9 May 1860 and was stored at the docks awaiting its dispatch to the Good Hope mine some 170 arduous miles over the Great Dividing Range.

In April 1851, the first payable gold in Australia was discovered at Ophir, approximately 28km from Orange. A number of explorers had travelled through the area known as Blackman's Swamp in the in the early 1800s, including Surveyor-General Oxley in 1817 and 1818. The area had been settled in the 1820s and a survey of the district, by J.B. Richards began in 1828. By the late 1830s, land clearing and cultivation had begun by farmers and by 1845 a village had begun being developed at Summer Hill to the north-east of Orange, at the junction of Gosling and Fredrick's Valley Creeks. It is thought explorer Major Thomas Mitchell proclaimed the village Orange on 18 November 1846 in honour of Prince William of Orange. Orange village had only around 28 people before the gold rush and although thousands of people flocked to the district's diggings, including Lucknow, the population of Orange had only grown to around 600 by 1861. An inn named The Bush had been opened and a shoemaker, two stores and a tannery were in operation as well as the earliest industry, flour milling. A slab-and-bark hut served as a Court House and was also used as a church by visiting clergy. In 1858, a court was established at Orange, and the township became a municipality on 9 January 1860. In June 1862, Cobb & Co. established its headquarters at Bathurst and the following month coaches were passing through Orange to and from Forbes to the west.

A few miles to the south, on the western side of Cadiangullong Creek, Saul Samuel, Randolph Want, Thomas Icely and John Savery Rodd began planning the sinking and driving of shafts on various areas of a lode which had been exposed on their 565 acre property. Under the name Canobolas Mining Company, they had issued a prospectus in September 1859 to raise £30,000 that would allow them to continue development of the mine. Captain Clymo, who had visited the mine earlier in March, wrote glowingly of its richness:

> ... it is one of the most magnificent deposits of copper ores ever found in any age or country. I know most of the richest mines in Cornwall and Devon; I have also seen the celebrated Parry's mine in Anglesey, and am persuaded that this will rank second to none of them; indeed it is equal, if not superior, to the Burra Burra itself.

Open-pit mining on a small scale had commenced, and over 250 tons of ore extracted yielding 30-35 percent copper. In writing to Josiah Holman, John Christoe described the lode as being a chain (20 metres) wide and very rich in copper. In late 1860, seven tons were shipped to Sydney and ingots of *"pure metal of the highest quality"* were exhibited at Saul Samuel's Pitt Street offices.

Having returned from London in late 1859, Morehead was cautiously looking for new, more profitable mining opportunities. In July 1861, SAMC took a 21 year lease on an area of 564 acres on the eastern side of Cadiangullong Creek, 15 miles south of Orange. Morehead had negotiated a royalty payment of one twelfth (8.3%) of the value of ore raised. Operations began almost immediately at their Oakey Creek Copper mine which was later named the Cadiangullong Copper mine. Morehead was confident in John Christoe's ability to plan the development of a smelting operation for the mine and other nearby mining companies and engaged him prior to mining operations beginning in July. It is likely Morehead negotiated a smelting agreement with Saul Samuel prior to commencement, given its proximity and recognising the huge cost of cartage.

Christoe directed the men to begin excavating a large shaft 10 ft by 8 ft sunk to a depth of 18 ft in the south section. Success was almost immediate when a rich lode was struck some 70 ft wide. Blasting and further excavation continued a further 10 ft confirming that the richness of the lode continued at least that far. Exaggerating somewhat, Christoe was of the opinion much of the ore was 40 percent copper. Water from small springs and a wet winter plagued progress and the lack of a competent smelting workforce impacted progress at Cadiangullong. Although joined by Thomas Hussey, a smelter whom Christoe had worked with before at Byng, Christoe recommend the company take steps to bring experienced smelters from Wales. Morehead quickly recognised that the additional workforce would require more permanent accommodation than the tents dotted about the hillside. A building force of over 50 men and families were employed and sent to Cadia to build slab huts and stores as well as assist with the development of the smelting works. The company engaged additional miners at the rate of 22 shillings per cubic fathom. By mid-September, the hillside shaft had reached 32 ft and it had been determined by the use of various drives and adits that the lode extended at least half a mile through the property and was mostly 70 feet wide. Although water continued to plague the mining operation, by October they had extracted over 600 of tons of ore which Christoe determined, more correctly this time, to be 12 to 15 percent copper.

In late 1861, at the Good Hope mine near Yass, now under the direction of Captain Perry, expectations were high in anticipation of rich rewards when the copper lode was discovered past the 180 feet level. A cross-cut which penetrated the lode for 5 feet was described as containing "blue spa combined with a large quantity of yellow sulphuret of copper and little mundic".

In December, Morehead travelled to Yass to visit the mine, only to find it had filled with water. All there was to see were deep pools where the shafts began. Disappointed, he then travelled to Cadia and visited Christoe at the Cadiangullong mine. Christoe had directed six men to excavate a cross-cut through the lode. The rest of the men were engaged in development of the smelting works, including a water-carrying culvert which had already been lined for a considerable distance. Hundreds of tons of wood had been cut and stored nearby in anticipation of the completion of the furnaces early the following year. Morehead examined the lode along the south shaft which extended deep into the hillside and was impressed by its thickness and its red, yellow composition. He also took time examining the near completed smelting works, the chimneystack now reaching 75 ft. One of the furnaces was already operational with two more under way. The disappointment of Yass was overshadowed by the progress and successes he observed at Cadia.

Returning to Sydney, and based on his recent visit, Morehead decided to send the large 25 inch Thomas engine and boiler to Cadia. Realising the need for a very experienced and competent mining manager to design the layout and install the large machinery, he contacted Christoe and requested he write to Captain Holman. He also placed an advertisement in the Sydney newspapers seeking new or second-hand water pumps.

With the smelting works almost complete and an improvement in the weather, the focus turned to further development of the mine. The Engine shaft had reached 40 ft and Murphy's shaft to 55 ft which enabled Christoe to confirm the lode was predominantly mineral-producing. From these excavations, 100 tons of ore was raised making the total over 800 tons at grass waiting for the smelter to be completed. In Sydney, Morehead advertised for two steam boiler makers to fabricate a large boiler. There was much celebration at Cadia when the 25 inch rotative beam engine and machinery inched its way over the Great Divide and finally arrived. This was possibly the largest piece of heavy equipment to cross the Blue Mountains at that time. The cost of transfer was estimated by some to be over £1500.

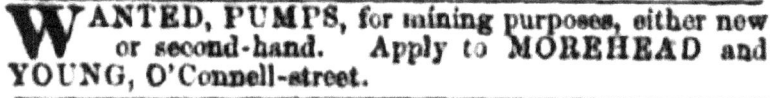

The Sydney Morning Herald Monday 20 January 1862

Meanwhile, in South Australia, the Bon Accord mine was also progressing slowly as Captain Dalley's recommendation to explore further areas to the east of all workings had yet to be undertaken. Also under consideration, and perhaps delaying progress, was a suggestion put forward by Elljah Whitford and a number of miners previously working the Burra mines nearby. To "prove" the lode, they contended, a shallow trial to the west of the current workings should be undertaken.

Captain Holman Arrives
1862-68

Reaching Sydney after nine days at sea, Josiah was greeted by a bustling port. Large ships, 1000 tons or more, were anchored in the deep, calm harbour, surrounded by sandstone escarpments. Houses dotted the horizon between rugged cliffs and, further on, clusters of impressive buildings loomed —many built of chiselled sandstone. Smaller boats ran to and from the ships. Ferries looped from bay to bay, picking up passengers at various wharfs. The streets were surprisingly clean, despite diverse building works that blotted the landscape. Development was going on at a pace in the prosperous colony of New South Wales.

Once ashore, Josiah found his way to O'Connell Street and Robert Morehead's office where he was greeted by Mathew Young, as Morehead was unavailable. Young took him to the Tattersalls Hotel in Pitt Street where he would be staying for the next two nights. That evening, Josiah met with Robert Morehead. There was a certain affinity between them, both having been uprooted from their family, friends and a certain level of civility they had enjoyed in years gone by. During their meetings, Morehead told of his almost two decades in the settlement and his first-hand knowledge of the growth of the young colony. For Josiah, this provided great background to what was to transpire in his new venture.

With his valise, small mining equipment and trunk, he caught a cab to Central Station where he boarded a train to Rooty Hills to begin his long journey by coach to Orange. While planning had begun for a rail line across the challenging Blue Mountains, Orange was then five days distant by coach.

Uncomfortable as the journey was, Josiah enjoyed the vista over the mountain ranges and descent onto the rolling, open plains. He reached Cadia on 25 March. Travelling over the crest and edging down the steep decline, he could see a village in the distant valley. The rolling hills reminded him of Cornwall, although much more densely forested around the steep hilltops. The site itself seemed rather primitive — rows of tents, half a dozen houses completed or partially completed, an assay office and smelting shed, manager's house and large captain's house and office. There was of course a blacksmith and carpenter shop and a number of store sheds along the flat portion of the valley.

Oakey Creek, as it was known by locals, offered fresh water from springs in the hills above the village and divided the settlement into two parts; a bridge of sorts connecting them. Rows and rows of huge bright-orange Silky Oak trees lined the stream banks. Miners' huts and tents could be seen scattered about somewhat indiscriminately in the shallow valleys of the two hills. On one side, in the distance the Iron Duke open cut mine was bathed in sunshine, a dark shadow indicating a shaft entrance. Immense piles of ore from the shafts dotted the landscape, dispersed sporadically over various sites nearby. Near the mine, oxen could be seen working hard raising the ore in large iron buckets to be added to the ever-increasing heaps. The streets were in poor condition but some makeshift bridges had been added to improve movement between the mining area and small township. The operational area comprised blacksmith's and carpenter's workshops and a number of small storage sheds just below the chaplain's house. Just to the north, timber had been stacked, much of it sawn to size for the furnace and on the level areas, bricks were stacked and stockyards were being added. Crossing Cadiangullong Creek, slab huts in the midst of construction lined the street in between the village store and post office, bakery, butcher and Bon Accord hotel. Where level space allowed, tents filled most of the vacant land. The smelting works under the supervision of Christoe was nearing completion on the southern end of the property and on the eastern bank of Cadiangullong creek, a site was being cleared and foundations pegged out for the engine house.

After finding John Christoe and catching up with progress, Josiah wasted no time reviewing the operations of mining activity and began planning the next phases of the mine development. Continuing the south shaft had already yielded over six tons of ore with an estimated 10 percent copper. Checking the various heaps of ore scattered about, he determined there were perhaps 1000 tons of ore at grass of which about half were 12 to 15 percent copper. There were certainly some very rich ore specimens, up to 45 percent but probably only 100 tons, the rest were less than 8 percent;

Vibrant "peacock" ore of many coloured pyrites displaying rich and intense green, variegated with red and yellow carbonates intermixed with brilliant sparkling quartz could be seen.

Having completed his assessment, Josiah directed eight miners to begin the stoping of the older engine shaft, others were directed to continue digging the water race. Miners engaged in stoping, wheeling or removing, the overburden were paid one shilling per cubic fathom. A few miners were directed to excavate a crosscut north and south for a length of 15 fathoms. It was while a winze was being dug for ventilation that the miners exposed a very rich lode almost 9 ft thick, which they estimated to be over 25 percent copper. As the miners continued Murphys shaft, at the 10 fathom level the lode quality improved considerably. In his report, Josiah wrote ... *The great width that quality of the ores rival the celebrated Springbok Mine in South Africa, judging from what I saw of it in 1856.*

Josiah, after careful inspection of the new mining equipment, determined it was in good working order. Christoe had completed the culverts for the furnaces, ordered fire bricks and quarried stone in preparation for the arrival of the masons. One of the older shafts carved into the hillside had been expanded and would operate as a forge. Josiah meanwhile directed a number of men to begin cutting out the foundations for the engine house on the southern mine area, while three teams were sent out to cut and cart wood, Box and Red Gum, to be stacked nearby.

With the smelting works and two furnaces completed, the first test was undertaken on 21 May 1862. A number of carts loaded with ore waited on the tramway above the furnace ready to be moved and emptied into the furnace. A single furnace was able to smelter about two tons of mineral ore. A number of experienced Welsh smelters lingered around waiting in eager anticipation, their long rakes at the ready to stir the heated metal and rake off any refuse. Huge large blackened gloves of leather and thick canvas type aprons lay nearby, all showing signs of wear from being scorched and singed by the intense heat of the furnace. Pigs (ingot moulds) lay about, ready to be filled with the rich molten copper. Each would weigh about 25 pounds when filled. As the heat within the furnaces increased, a cracking noise could be heard emanating from within the furnace. One by one, the bricks used to line the furnaces began to collapse. This was not a good start and it was unfortunate timing as Morehead and a number of investors happened to be visiting when the smelting operation failed. After conferring with John Christoe, Josiah ordered a large quantity of firebricks from Campbell's River south of Bathurst in the hope these would survive the intense heat. The disappointment of the collapse of the furnace was overshadowed shortly after by the arrival of John Christoe's father-in-law, Dr Matthew Henry Smyth Blood and family from South Australia. Dr Blood had been engaged by SAIC as a general practitioner (GP) and medical superintendent, as by now, almost 200 men were employed in various capacities at the Cadiangullong mine.

Further assaying in July revealed the high-quality ore contained much more iron than previously thought and the percentage of copper was less than 24 percent. Josiah directed the men to extend the stopes on the ore floor, in the hope of determining if the rich lode already found extended both at depth and length. With the furnaces now working, Christoe was pleased that the ore, now with the right quantity and mix of flux, smelted "admirably well". He also learnt that the Welsh smelters and miners had arrived in Sydney and were heading west, a number having honed their smelting skills at Carmarthen, Wales. Some would do well in the colony, like Lewis Lloyd who would work in a number of mines before purchasing his own and becoming one of the richest men in New South Wales. Others returned home, missing family and friends or finding the harsh unyielding country far too demanding.

The vagaries of the weather continued to plague mining operations at Cadia. A wet winter with snow swelled the creeks often making travel perilous. Roads, when usable, were boggy, thus preventing haulage of heavy loads of ore and wood. The mine also suffered, as the expected excavation of 100 tons of ore per month could not be achieved in such conditions. As well, the Welsh smelters, accustomed to using coal, encountered a different, more demanding challenge when using wood. Added to that, with all the rain, trying to keep the wood dry became a constant battle. The deeper mine shafts were continually being filled with water, the small pumps having to work day and night. By the end of spring, with clearer skies, most of the winter obstacles had been overcome. From the heaps of ore at grass, 13 tons of fine copper was on its way to Sydney with 10 tons ready to be dispatched as soon as a carter could be found. When further extending the engine shaft northward, at 23 fathoms water poured in and it was not until an additional engine and pump was purchased for £350 and sent from Sydney, that they were able to continue. Specimens of ore were obtained at this level of the shaft and were sent to the Sydney Mint to be assayed. Chemist Mr Watt and Professor Smith of Sydney University examined the ore and discovered that gold was intermixed with the ore. This was indeed an exciting discovery and vindication of Morehead's decision in the mine investment. But mining is not without risks and there are always likely to be accidents of one type or another and on the afternoon of Sunday 7 December while the miners were having the day off, not starting until 10 p.m., a fire broke out in one of the sheds near the No 2 furnace. It took far longer than normal before the men were assembled to put out the fire and by then over 200 tons, almost a quarter of their stored firewood, was destroyed.

Frustrated by lack of progress at the Bon Accord mine in South Australia, Robert Morehead and daughter boarded the 700 ton *City of Sydney* on 30 October 1862 bound for Melbourne, arriving two days later. On 5 November, they boarded the 330 tons steamer *Havilah* for Adelaide, arriving on the eighth, then took a coach to the Bon Accord mines. There they met Josiah and spent a number of days examining development of the works and the new shafts. Morehead spent almost a month at the Bon Accord mines, and on 13 December boarded the 531 tons steamer, *Balclutha* for Melbourne and then the 700 ton *Wonga Wonga* for Sydney, arriving on December 20 in time for Christmas. Josiah returned and stopped at Yass to visit the Good Hope mine to review progress. He judged the lode to be around 12 ½ ft wide and at the 180 ft level, determined the yield to be around 10–12 percent copper. He observed a second lode adjacent to the original lode at this level and assessed the mine could be worked down to 300 ft. He also suggested, as the Yass River was nearby, that its water might be used to drive a waterwheel to pump out water that was continually plaguing the mine.

In mid-January 1863, the first shipment of copper ingots from the Cadiangullong mine, around 12 tons, left Sydney docks for London fetching around £92/50s per ton. The water pumps, having *answered its purposes satisfactorily*, allowed Josiah to at last examine the engine shaft and instructed the men to blast a further 10 fathoms through the hard ironstone. He also directed a team to drive a crosscut south and north 7 ½ fathoms and west 10 fathoms through the lode in line with Murphy's shaft. By now, Christoe had the smelting works running extremely efficiently and was smelting ore from Cadiangullong, Canobolas and other mines in the district. In March, a further 40 tons of copper was sent to Sydney. Having excavated the original shaft a further to 13 fathoms, Josiah determined that the prospects of the Cadiangullong mine (East Cadia) were no longer tenable. Added to that, continual water entering the mine was creating great difficulty for the miners. Any thought to extend the mine further would require the purchase and installation of a larger engine to pump out the water. In his report, Josiah wrote ... *inclined to think that the prospect of the mine were not such as would warrant him recommending the outlay which this would involve.*

In March 1863, Morehead, on the advice of Captain Holman, suspended all work on the engine house building at Cadia until the results of further exploration were completed. Since his arrival, Josiah had been exploring nearby areas showing potential and exhibiting positive features that would warrant trials. With work on the engine house now suspended, he directed a small team to carry out a number of shallow excavations in a few of these areas on the opposite side of the creek known as south section and north section. Where quality ore was discovered, additional shafts were sunk to a depth of 17 fathoms.

Meanwhile, in South Australia, at the Bon Accord mine, Captain Dalley had reached depths of 24 fathoms on the eastern side of the claim without finding any ore of quality. Further to the east, another mining captain with 10 years' experience, Captain Simmons was instructed to sink two shafts to the 80 fathom water level. Still not finding the continuation of the very rich Burra Burra lode from the adjourning property, Captain Simmons suggested continuing below the water table. Asked by Morehead for his opinion, Josiah remarked: –

> ... *sufficient amount of exploration has been made in sinking shafts and cross-cutting, to prove that no large body of ore exists above the water level in the Bon Accord property, in the locality where the lodes of the Burra Burra property should have been met within the Bon Accord property, if they had been continuous from the former through the latter.*

In April 1863, at the general meeting of SAMC held on the 22nd in London, a directive was sent to the management committee in Adelaide notifying them of a resolution that had been passed to resume operations at the Bon Accord mine at a greater depth.

At Cadia, Christoe's smelting works were running exceedingly well and in April, over 200 tons of excavated ore was delivered for smelting. Almost seven tons of rough copper was waiting to be refined and 11 tons of fine copper was ready to be transported to Sydney following the 17 ½ tons already dispatched earlier in the month. On 15 May 1863, Morehead finalised arrangements with the landowners for a further 12-month trial period which would include the former Canobolas mine. On the west bank of Cadiangullong Creek, where Josiah had carried out trial excavations, about half a mile to the south of the original shaft a new mine was developed called West Cadia, but later named White or Big Engine Mine.

As the 12-month trial had "proved" the prospects of the West Cadia properties, all the focus was now directed to that mine, to the detriment of the Canobolas mine. All was not lost however, as abundant sulphide to aid smelting had been discovered at the Canobolas mine. Josiah spent some time mapping out a series of adits or tunnels in the newly discovered West Cadia mine areas which would, if necessary, allow drainage of the mine at very considerable depths and permit the ore to be raised very cheaply. Additional shafts continued to be constructed west; Lawson's shaft, 25 fathoms deep, and beyond that Want's shaft, also 25 fathoms. A drive of 64 fathoms in length following the course of the lode was completed to connect these shafts.

East of the Engine shaft, Martin's, Baker's, Clarke's and three other shafts were planned, the deepest of which would be 15 fathoms. A drive was to connect the first two of these shafts at the 15-fathom level. At various levels, over 750 fathoms or nearly a mile of mine driving was undertaken supported by sleepers, runners and curbs. There were 13 shafts in total, three of which were 40 fathoms and 10 varying from 10 to 25 fathoms. The Big Engine shaft, the largest, was 12 ft by 5 ft, another 9 ft by 4ft, each with two chambers. The smallest shafts were 6 ft by 4 ft and all were slabbed. A series of tramways was planned that would carry the ore to the surface and beyond.

Another wet winter again thwarted their efforts to raise the intended 100 tons of ore a month after only 40 tons a month was excavated from the various shafts, averaging a yield of 12 percent copper. By October, blasting at the Cadiangullong mine had begun in earnest and 20 tons of ore was raised, some samples as much as 20 percent copper. When the excavated and crushed rock was removed in the Canobolas mine, stoping re-commenced and Josiah was able to determine the estimated size and value of the rich lode. The Canobolas mine produced almost 26 tons and from West Cadia Clarke's, Want's, Rodds, Halls and Samuel shafts almost 84 tons averaging 11 3/8 percent copper. On an even more positive note, prices of fine copper in London remained strong at around £100 per ton. The smelting works had 22 tons of fine copper waiting to be transported to Sydney and the fine weather had allowed them to increase their stocks of wood to over 3500 tons. Richard Morehead visited Cadia and reported:

West Cadia mine prospects on the whole, encouraging as respects the prospects of a permanent mine.

Of the smelting works, he was encouraged by Christoe's confidence he could send 30 tons of copper ingots to Sydney by the end of October and thereafter, 20 tons each month.

However, patience had run out at the Bon Accord mine in South Australia; when reaching a depth of 100 fathoms without finding ore of quality the chairman of the committee wrote: "... I cannot recommend further expenditure in improving it . . I intend to visit the mine this week, to see that everything is left in order, and by next mail we shall be able to forward inventories and close accounts of expenditure on the mine:"

Within six months, mining had ceased and an offer was made to take over the mine by the Yorke Peninsula Mining Company Limited.

In December, Josiah reported that 120 fathoms were excavated from the lode at West Cadia mine, the yield estimated at around 100 tons of 12 percent Copper.

He wrote, *I estimate the next sampling for West Cadia will rather exceed the past quantity and of average quality, whilst the cost will be less, and a considerable increase in the yield of ores may be looked for from the Canobolas mine, with little or no addition to the costs.*

In May 1864, with the very favourable prices of quality copper being maintained in London and the rail link completed to Penrith, which would significantly reduce cartage costs, Morehead made an offer to purchase Icely, Jones, Rodd, Samuel and Want's Cadiangullong Consolidated Copper Company. It was decided to merge the two operations and ownership and establish a new entity, the Cadiangullong Consolidated Copper Mining Company. A notice appeared in Sydney newspapers announcing that:

A company has been formed called: –

The Cadiangullong Consolidated Copper Mining Company, for the purchase and working of the two mines known as "East Cadiangullong" and "West Cadia".

The property consists of 1500 acres of freehold land and is situated about twelve miles from the town of Orange. The smelting works are in full operation, and nearly 200 tons of fine copper have been produced since they have been erected. The capital of the company is £60,000 in 60,000 shares of £1 each, the whole of which have been taken up with the exception of 10,000 shares (which are not to be issued at present).

Thanks to drier summer months, the mining operations now progressed more smoothly and the village of Cadia flourished with that success. A mail service ran past the village to Carcoar, Stepney Clarke's store was operating as the acting postal agent and there was a telegraphic wire service at Orange, just 12 miles away. The mining captain's house had been extended, now a substantial 60 ft by 30 ft and new shops appeared as the population had by now almost doubled. The target of 100 tons of ore raised each month was easily surpassed keeping the smelting works operational at full capacity.

A reporter visiting Cadia wrote …

The country is heavy, mountainous, and broken, cut up by numerous creeks and water courses, which are sometimes hemmed in on either side by steep banks of schistose rocks; at other times they open out into beautiful level flats, or gently undulating rises that blend gradually with the ranges that tower above them. It is on such a piece of gently undulating land, having a broad flat at one extremity, that the mining settlement has been formed. A fine creek of bright and sparkling water, fresh from the springs of the giant hills above, winds its way through the spot, dividing the settlement into two parts.

On either side of the stream, the huts of the miners are scattered about in all directions, now nestling down in the dip between two hills, now perched up on some bold rise, or again rising soberly from the level area. Stores and shops, in different lines of business, are ranged in something approaching to a row, whilst trades of various descriptions — the butcher, the baker, the shoemaker, the tailor —-are carried on as actively as if the work was not being done in a mountain fastness.

A post-office has been established on the settlement, the name having been very wisely abbreviated into Cadia, a title that no doubt causes some of our Sydney residents who see it in the Post-office list, to wonder where on earth the place can be, since the name, though gradually getting into use in the immediate district, is utterly unknown beyond it. Cadia is almost on the direct line between Orange and Carcoar, and a mail runs through once a week; but the driver has to pass on sufferance through several private properties, there being no proclaimed road either between Cadia and Orange, or between the former town and Carcoar. This is looked upon as a grievance, and with great reason, the more especially as a number of free selectors have taken up land along the line between the former places, and considerably complicated the road. As this is a matter of great importance, not merely to the full development of the work carried on at the mines, and to the free selectors settled in the neighbourhood, but also to the district at large, I have no doubt but that the attention of the Government will be given to the subject.

By mid-1864, almost 1000 tons of ore had been brought to the surface over the previous 12 months from the West Cadiangullong mine. Seven tons of fine copper from the newly formed Cadiangullong Consolidated was sent to Sydney and auctioned by Messrs, Mort and Co., selling for £92/5s per ton. It was hoped when the extended tramways were completed and additional train carts procured, the expensive procedure of raising ore via winzes and buckets could finally come to an end.

But what of wife Elizabeth and his six children far away in New Zealand? With confidence in the mine now well established by the new listing and capital raised, Josiah wrote to Elizabeth in Whangarei. Having decided this may well be his last adventure, Josiah applied for and was granted Portion 150 within the Parish of Waldegrave, around 100 acres along Cadiangullong Creek with the intention of building a home about a quarter of a mile south of Cadia village. He later applied for Portion 159, 120 acres adjoining his property to the south.

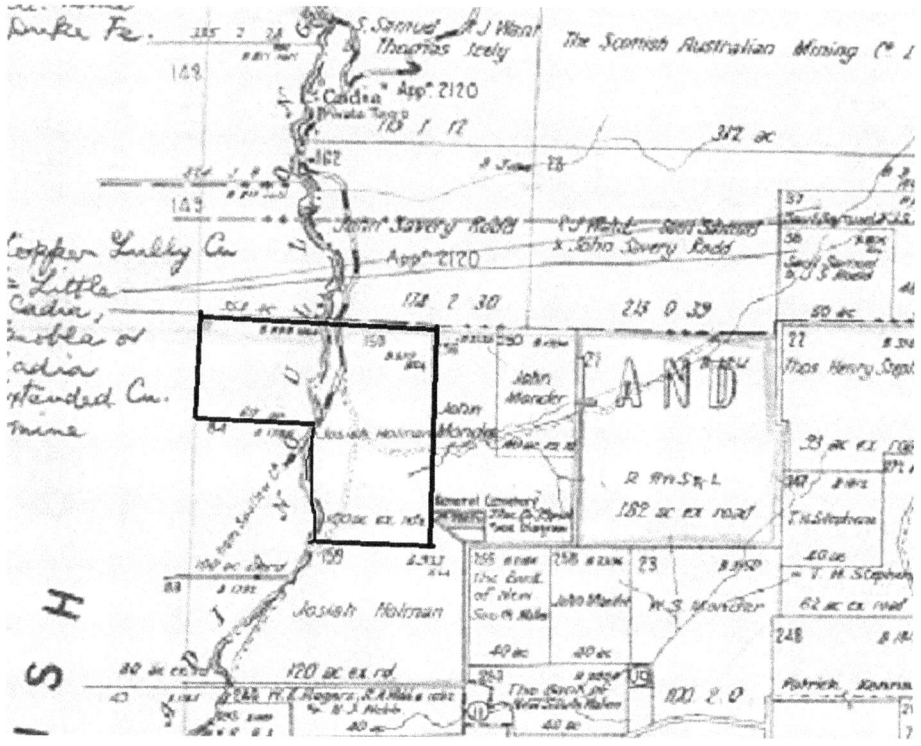

Section of Parish of Waldegrave map showing Josiah Holman Portion 150

He had earlier made enquiries about land purchase, only to discover that the company owned the land and miners and their families were excluded from purchasing allotments for their own homes. While shopkeepers and others could take out leases on which to erect their own buildings, freehold title was only available beyond the mine boundaries. To a certain extent this suited many of the miners, realising that should Cadia mine come to an end, they would have to search for other opportunities anyway and would not therefore be tied by leases. It did however lead to a restless life pitted against the needs of the family for a stable home and education.

Early large-scale pastoral settlement of New South Wales had effectively excluded small settlers from owning land until the Robertson Land Acts (Crown Land Alienation Act) of 1861. When squatter, Gustavus Richard Glasson, became aware of this, he asked that land in the area be put up for lease by auction and successfully bid for the lease. Others, including Josiah

Holman and Dr Mathew Blood would follow, applying for various available lots. The leasehold laws also threatened to deprive the Cadia community of feed for the teamsters' horses and bullocks, as well as timber for fuel, most important for the smelting operation. In 1866, Cadia residents and miners successfully petitioned for a Common to be declared around the mine to reserve these resources, but a further consequence of the Common was that it restricted Conditional Purchases in that area. This persisted until 1879 when the Common was revoked.

With much of the preparatory work for the large engine house completed and additional funds available for mining equipment, excavation of additional shafts began in earnest. Additional tributers were engaged; the building of a fifth furnace had been almost completed and planning begun for a sixth furnace as well as erecting machinery for dressing the ore. It was the mining captain's responsibility to ensure the profitability of the mine, his ability to determine the likelihood of which areas underground would contain the richest lodes rested on his shoulders. To do this, he would engage tributers — experienced hard rock miners — to work the ore face. Once the underground shafts and drives had been completed, the mining captain would examine these and select areas known as "pitches" to be let out for tribute. This Cornish custom began at the beginning of the year, and each ninth Sunday was known as Survey Day. The skilled captain having determined the likelihood of rich ore in the "pitch", would offer it at a price. How well he determined that price would have a bearing on the profitability of the mine, it required careful balancing. As the captain called out the number of the "pitch", he would offer a price and if there were no bids, he may well lower the price until accepted by one of the tributers, who generally worked in teams. If the tributer, who had earlier examined the "pitch", suspected it was rich in minerals, he may well offer more than that determined by the captain, the pitch then would then be allocated to the highest bidder. Settlement day was the day before survey day when the men would be paid based on the yield from the ore they had mined in their "pitch".

Much of Josiah's attention was focused on planning the construction of the "Cornish type" engine house on the southern section of the mine site. Using local stone for the first 50 ft and locally manufactured brick for the 16 ft chimney, it was to house the powerful 40 hp beam engine which would drive the rotating fly and driving wheels which were 27 ft in diameter. These revolving wheels would drive the 7 inch Cornish pump and crushing machine as well as power other machinery for jigging, sawing and hauling ore from shafts as deep as 38 fathoms. Attached were powerful self-feeding, regulating rollers to break the rock sufficiently to pass through the regulated grates. This resulted in very efficient smelting using the furnaces with connecting channels dug into the hillside to direct water to the site for ore washing. The engine was powered by a 10 ton Cornish boiler, 7 ft in diameter and 31 ft long with a flue 4 ft in diameter, one of the largest in the colony. The most unusual feature was the strong outer shell which was composed of 13 conical strakes to make up the 31 ft length, each strake having 5 plates to form the 22 ft circumference, each plate a nominal 4.5 ft long by 2.5 ft wide. The boiler made of wrought iron was secured using hand-closed rivets. A single gusset stay at each end supported the flat end plate above the furnace tube.

The assay building with assay furnace, crucibles and all necessary apparatus for assaying was attached to the engine house, where samples were taken from the ore as it was mined. When crushed, the ore was analysed to determine the amount of copper that it contained. By any criteria, this was a major undertaking using all the skills Josiah and John Christoe had learnt over the past decades. Josiah ran a tight ship, the evidence of which can be seen in the original ledgers that still exist today and are stored in the Millthorpe Golden Memories Museum. A milder winter allowed for greater production and by August almost 300 tons of ore had been smelted yielding an average 12 percent copper with over 400 tons at grass. Eighteen tons of refined copper had been sent to Sydney with 25 tons being loaded for transportation.

In September, Josiah wrote:

The nature of the rock surrounding the lode in the East shaft looks very congenial for copper ores, and I am sanguine that the lode will be found, when opened out in the adit level, more productive than anything exposed from sinking the shaft.

After spending Christmas with John Christoe, his father-in-law Dr Blood and family, Josiah caught the mail coach to Orange and then to Penrith to catch the train to Sydney. He took the opportunity to meet with Robert Morehead and on Wednesday, 11 January 1865, waited at the docks for the *Egmont* to arrive. It was a very happy family reunion, the children overexcited to be reunited with their father after three years' absence. Up early the next morning they caught the 6 o'clock train from Central Station, stopping at Newtown, Petersham, Ashfield, Burwood, Homebush, Haslem Creek, Paramatta where they changed trains and joined the Great Western Line to Blacktown, South Creek and finally Penrith, arriving just before 8.30. From here they boarded a Cobb and Co., coach, £2/5/- per person to Blackheath staying the night at the coach inn and then to Bathurst where they stayed at the Elephant and Castle Hotel. The next morning, they continued onto Orange where they bought provisions and rested for the evening at the Royal Hotel. Very early next day the family headed out with much excitement and anticipation along a not so comfortable bumpy road, crossing numerous small creeks on the way to Cadia, just a few miles south. Arriving at the captain's house provided by the company, Elizabeth and the children began to organise and arrange furniture to their liking while Josiah checked the latest status of the mining operation.

On the 28 February 1865, the Cadiangullong Consolidated Copper Mining Company Limited held their first general meeting of the shareholders. They reported that expenditure to date was over £12,000 and that 830 tons of ore had been raised and dressed producing 106 tons of high-grade copper. Of this, 51 tons was sold for £4429, the remaining estimated to realise £4733 and taking into account over 300 tons of dressed ore was at grass still waiting to be crushed and jiggled, the total would more than cover the cost of expenditure to date. The smelting works comprised four ore and roasting furnaces, one of them erected since the establishment came into the hands of the company, with two additional ore and roasting furnaces almost completed. This included the building of the calcining furnace which had recently been completed so the more sulphurous ores could undergo a preliminary process prior to final smelting. No dividend was declared. Record copper prices had held for almost 12 months, but this could not go on for ever and in September 1865, Mort & Co., reported selling 55 tons at £82 10s per ton.

With the engine house stonework now completed and roofed in, shipments of fine copper left for Sydney almost fortnightly. At sea on the way to London in mid-1865 were 15 ½ tons of fine copper, and waiting on the docks for shipment was 30 tons with a further 22 tons on the road to Sydney. Under the oversight of Holman and Christoe, the village had now grown to over 500. Stores and shops hosting different lines of business had sprung up along the meandering street. There were three stores, the largest owned by William Blood, and two Inns of sorts — built in the typical style of slab huts with two rooms, a bar and two adjoining dormitories offering the most basic of accommodation. But it was predominantly a mining town, and its growth attracted elements of undesirables especially as the nearest police station was 12 miles away at Orange. There was a fine stone church of course, for those wishing to repent. A mail delivery coach ran each week from Orange to Carcoar for miners' families and pastoralists and the district's free selectors.

The Cadiangullong Store, owned by Stepney Alured Clarke, had operated an unofficial mail service since 1861, but in October 1863 Holman and Christoe petitioned the State Government to establish an official post office. Christoe was appointed the first postmaster in July 1864. On 1 August 1864, storekeeper Stepney Clarke, who had run the unofficial post office for over three years, applied for the position of post master and was appointed on 1 September 1864. Storekeepers always tried to secure the post office, as its presence was a stimulus to other business.

After the establishment of the post office, a public school was the next priority for the growing village. On 31 December 1863, Christoe prepared a school application stating around 65 to 90 children from mine families and about 25 from surrounding farms could attend the school. A public meeting had elected a representative sample of residents as patrons, included John Christoe, mine manager (Church of England), who was elected as secretary, Josiah Holman, mining captain (Church of England), William Blood, accountant (Church of England) along with others. Approval of their application would certainly have been assisted by having Robert Morehead on the National Education Board. Approval for the local National School was received in early 1865 and a fine brick building was constructed, furniture ordered, all hopefully ready for the first term of 1866. The Cadia National School opened in December 1865 with Henry Bonnar as its first schoolmaster. Josiah's children, Emily 16, Josiah Jnr. 13, Charles 9 and Annie 6 attended the school with the other children, in total 12 boys and 8 girls. Annie would later write:

> "I was born on November 2 1859 and can remember starting school at Cadia when six years of age." **Central Western Daily,** *May 11 1950.*

In the bustling village, miners trudged off to their mines, sinking shafts, blasting and picking rock to be loaded and taken to the top, sometimes working at depths over 200 ft, although these were far less than the half mile long copper mines of Cornwall. Older boys often assisted, sorting through the smaller pieces of ore, selecting those that may have a metal vein running through them. A travelling reporter described the process this way:

> The first process in treating the stone is done by boys, who handpick it, separating the ore from barren rock, the ore being removed at once to the furnace, the intermediate or coarse ore only going through the processes of crushing and jigging. Appleton's stone-breaker driven by a belt, begins by cracking the stone neatly to the size of road metal, whence the metal

is removed to a hopper, beneath which it meets with heavy Cornish cast-iron rollers, which again reduce and discharge it in a wheel fourteen feet in diameter and ten on the rim, having internal buckets which carry up again the stone, which is still too large to pass through the meshes of a circular revolving sieve. Eight jigging machines jumping under water in as many slabbed pits separate by specific gravity the ore sand from the silica coming from the sieve, and boys skim the latter off periodically.

Families shopping and completing other activities had become used to the roar of the furnace, the crushing of rock and the metal jangling of the steam engines. A visiting reporter summed it up best:

> The snort of the steam, the clang of the engine, and the roar of the furnace mingled with the hum of human voices give a stranger impression than that conveyed by a newly opened gold field, for above all and pervading all there is an unmistakable sulphury odour, somewhat irritating at first to persons of delicately sensitive olfactory nerves.

Near the southern section, the Cornish engine of 40 hp had been installed to drive the pump work, saw wood, haul the stone, and crush and jig the copper ores. On the northern section were two engines, one of 12 hp used for pumping water, hauling ores and rubbish for the Trevenas shaft, the smaller one of 8 was available for other purposes when required. The six furnaces — three for ore and metal, one for roasting, one calcining and one refining the mining site — were now running with maximum efficiency. Josiah wrote that they now had the capacity to smelt and convert over 200 tons of ore a month into fine copper, around 28 tons. The other crucial element, of course, for all this was water, and while some water from the deep mining activity helped, a severe drought over winter and into summer "*retarded operations.*"

One of the directors visiting the site reported:

I have satisfaction in stating that we are very favourably impressed with the newly erected machinery which we then inspected. It was disappointing, at the same time, to find that the engine, from a short supply of water, could not be kept at anything like full work.

In December 1865, a *Sydney Morning Herald* reporter visited Cadia:

"We had the pleasure of visiting this interesting locality during the past week, and were courteously shown over the various works by the several gentlemen present in charge of the departments. The extent and value of the mining and smelting "plants" and machinery, the hum of a numerous population, and the vastly altered appearance of the locality from what it was some nine years age, could not fail to impress us with the fact that the real value of such a discovery in our district has been scarcely appreciated as it should be, and the miserable state of the numerous tracks, leading to the mines (roads there are none) confirm the belief that much valuable time, trade, and opportunity, have been lost to Orange, which if once taken up by rival localities may never be regained.

At the time of our visiting a portion of the mines, a discovery had just been made of a new vein of "yellow ore" predicted by Captain Holman (a specimen of which may be soon at this office), and which, in conjunction with the careful and skilful supervision of that gentleman proves that the mines are in excellent hands, and that the prospects are steadily improving. We were much struck with the appearance of the splendid condensing engine of forty horse power, manufactured in Cornwall, and which appears fully equal to the requirements of the principal mine, whilst two portable engines work the other mines contiguous. Extensive improvements are being added for the crushing and washing of the ores, which, saves time and labour in the future process of smelting. The short supply of water this season causes a series of works to be brought into play for saving that very necessary element, which in former years was super- abundant. The smelting works under the able management of J P Christoe, Esq , are in full operation, with active arrangements for further extension, and the whole place wears an air of business and prosperity, cheering in these dull times to behold."

"But touching the social condition of the locality we heartily wish we could speak in equal terms, still there are appearances of an improvement on the present state, by the almost immediate opening of the National School, a fine substantial brick building, nicely situated, forty feet by thirty two containing the requisite class and other rooms for the teachers, and on the opposite side of the creek, a neatly arranged Wesleyan Chapel, in which an attentive congregation of worshippers is frequently to be found and we sincerely trust that when the school is opened courses of lectures may be delivered by those whose time and acquirements qualify them, on subjects calculated to improve and elevate the minds of the people. An extensive circulating library, containing useful publications, such as Chambers's Miscellany, History, Biography, etc, would greatly tend to soften the hours of toil and labour by the solace of intellectual instruction in their leisure moments.

Police protection is also much needed, for in a scattered population of some six hundred souls there occasionally happens need for the strong arm of the law to assert its supremacy, and the consciousness that grievous offences have been overlooked by the paucity of police, gives violent and lawless characters opportunity for insult and outrage that should not be permitted by the Government. The supply of pure water in the bed of Oakey Creek is almost inexhaustible, and many of the miners show great diligence in gardening on its banks. The hills around are being gradually denuded of their timber for the purposes of raising steam, and for smelting, giving the landscape a vastly altered appearance to the time when the silence of the valleys was only broken by the yell of the blackfellow or the whistle of the shepherd."

On Wednesday 28 February 1866, the Cadiangullong Consolidated Copper Mining Company (Limited) held their half-yearly general meeting. Captain Holman tabled a report in which he confidently reassuring the company its operations would soon become profitable. A letter from him, dated the 21st instant, was also very encouraging, forecasting an increase in the production of ore in the West Cadia mine. Climate would always feature in the success, or otherwise of the Cadia mine. A wet winter made it difficult to gather and deliver wood for the furnaces and cart copper ingots over boggy roads and swollen creeks. A summer drought with less water decreased utilisation of the smelting works and increased the cost of feed for bullocks. The use of horses only increased cost further by almost 50 percent. *... and it has not been till within six weeks past, on account of the great drought, that the mine has afforded anything approaching a minimum supply of water to keep the engine at work in crushing these ores, and for dressing the same.*

It was reported in March that over 230 ton of ore was excavated that month from the West Cadia mine. Josiah wrote:

> *A fortnight since the men were brought back near to the shaft, and in crosscutting further south the main part of the load was found, it is large and well-defined, and in driving west therefrom it is yielding two tons per fathom of average quality pale yellow soffit of copper ore. Six men are driving on the load at £6 per fathom, and two men have now started to drive east with similar results. The price for stopping ranges from 40s to 50s per fathom, by 10 men in all.*

White engine house, Cadia 1866 by R. F. Blaker

With mining operational costs of between £1200 to £1300 each month and the return on the sale of copper, should they continue mining 200 tons of ore each month at around £2000, Captain Holman believed there was no reason that the mine would not continue well into the future. If the costs of the mining and smelting departments could be maintained and copper prices remained high, a healthy profit would be realised. The big "if" of course was the price of copper, over which he had no control. By mid-1866 over 250, mainly Welsh and Cornishmen were employed in the mines, boys, girls, engine drivers, firemen and smelters. The company had built slab cottages with shingled or bark roofs to rent to families at moderate rates of one shilling a week, the miners being paid 40 shillings, engine drivers 50 shillings and smelters 60 shillings. From their wages, miners paid a weekly subscription which allowed them to access free medical care. Josiah had also engaged a number of tributers who appeared to be satisfied with the results of their labours.

In July 1866, Josiah, received notice that his father James had passed away on 2 July. He was grateful for the presence of his family, especially Elizabeth, to support him. He knew of course it was inevitable, and that when he bade him farewell eight years ago, it was most likely be the last time he held his father in a warm embrace. Just as it was 18 months later when, just before Christmas, he received notification of his mother's passing on 18 November 1867. Years earlier he had written ... *I may never see them again on earth but I trust God we may meet in Heaven.*

Now, with all five furnaces running continuously, the issue of fuel became of paramount concern. During the earlier dry months, in anticipation of another wet winter, Josiah directed the men to cut and cart as much timber as they possibly could. Christoe wrote ...

> The wood on hand at works I judge, at least, to be 2200 tons, and 500 tons paid for in the bush — the total value £625. There are also, belonging to the woodcutters in the bush, over 2000 tons.

Fuel would always be a resource of concern. While the men could go out and cut timber in the wet winters, boggy roads prevented them from carting it back to the mine site. Finding timber on pastoralists' land often led to confrontation, but in part was later alleviated by the granting of Commons.

A Reporter noted on September 1866:

> I am sorry to say that the extensive works of this enterprising company has been very much retarded by the want of a sufficiency of firewood to keep their furnaces in full operation, which is caused by all the lands around the company's estate being let by Government to sheep holders whose stock eat off every part of particle of vegetation and prevent all cattle from being able to work or live about the mine and also retards the works from progressing so rapidly as they otherwise would. There is no doubt but if a proper representation were made to the Minister of lands, the government would grant the company several acres of land as a commonage.

In the *Government Gazette* of 26 October, the following notifications appeared:

> Messrs. Josiah Holman, Henry Bonnar, and John P. Christoe to be Trustees of the land at Cadia devoted to temporary commonage.

As well, there was competition for labour, pastoralists needing men for various agricultural pursuits, especially in the drier spring and summer months which best suited woodcutting. Although the tributers felt confident their endeavours would achieve rich rewards, the amount of ore raised began to decrease. The 200 ton plus a month promised dwindled to just over 170 tons, and soon after to less than 100 tons, although the yield remained the same around 12 percent. Hardly surprising, the amount of copper ore shipped to London slumped from the 70 tons a month earlier in the year to less than half that. Repairs to the machinery continued to create havoc to the smooth running of ore processing. Added to that were further costs of engaging additional miners, six for Mitchell and Rands shafts and 10 at Phillips shaft.

Captain Holman's reports however continued to be encouraging:

> *It gives me much pleasure to report that the load immediately improved, it is now yielding fully 2 tons per fathom of 20 percent yellow ores … and a sense of optimism for the future of the mine, … It now gives evidence that it would soon bring the mine into a prominent and remunerative position.*

On the 30 January 1867 the Scottish Australian Mining Company held their half yearly meeting at which a dividend of 7 ½ percent was declared. The board however were not convinced of the viability of the mine and asked for a full report from Captain Holman. The chairman reported:

"Captain Holman in Sydney thinks the odds decidedly in favour of West Cadia turning out an eventual success."

Ore mined from Cadia continue to decrease each month, soon to 86 tons, then as low as 65 tons impacted by another wet winter, shafts filling with water and flooded creeks making cartage of the heavy ingots of copper to Sydney impossible for weeks.

Christoe watched on in discouragement as the smelting operations output continued to decrease. Added to that, requests to address ongoing maintenance issues were ignored. In far north Queensland, at a recent mineral discovery, Peak Downs Copper mine was desperately seeking competent smelters. Christoe was aware a number of smelters had left Burra Burra for the Queensland mine and decided to apply for the vacant smelting manager position. Accepting the position as head assayer, he left Cadia in January 1867. His father-in-law, Dr Mathew Blood, had left Cadia the previous year and returned to Kapunda, where he was elected Mayor in December 1866.

> PEAK DOWNS COPPER MINING COMPANY.
> Wanted, a competent MANAGER of Smelting Works. Apply, by letter, with copies of testimonials or references, to the undersigned. Mort's-buildings.
> GEORGE SAIGHT, Secretary.

On April 24, Robert Morehead and his two daughters left for London via Melbourne and Galle on the 1007 ton *Avoca* (R.M.S.) under Captain Farquhar. Morehead did not arrive in time to attend the half yearly general meeting of shareholders to June 1867, the chairman Mr Alexander Young referring at some length to the company's interest in the Cadiangullong mine which he regarded as the unsatisfactory feature of the company's report. The directors were quickly concluding the anticipated rewards from this investment were unlikely to materialise. Also of concern was the continual deterioration of copper prices from around £100 per ton to less than £80 per ton. This was partially due to new discoveries world-wide creating an oversupply and also reduced demand due to the end of the American Civil War in 1865. Again, a dividend of 7½ percent was declared. By the end of the year, production at the smelting works had significantly contracted, and much of the mining effort was being carried out by the few tributers still attempting to eke out a living. Morehead, still in London, was unable to convince the directors to continue with the mine and on 20 January 1868 all mining ceased and in February the smelting works were closed after the copper in the furnace bottoms had been extracted. In March, the directors reported that almost 19 tons of fine copper had been shipped to London with about 30 tons yet to be dispatched from Cadia. Morehead returned on the *RMS Geelong* arriving in Sydney on 17 April 1868.

In London on Friday, 15 May, 1868, at the Scottish Australian Mining Company (Limited's) half yearly general meeting — i.e. to 30 December 1867 — it was reported that over 600 tons of fine copper had been extracted and sold by the Cadiangullong Consolidated Copper Mining Company (Limited) since the mining operation began. However, the directors felt that:

> The continued depression in the price of copper has rendered it impossible to make the returns equal the expense of working, notwithstanding that every economy has been practised.

A dividend of 8 percent was declared.

In late May 1868, a notice appeared in the Sydney newspapers advertising an auction to be held on 16 June of all plant and machinery at the Cadia mine. The auction failed to receive a bid and the property was put up for sale. On Tuesday, 24 November, 1868, the Scottish Australian Mining Company (Limited) held their half-yearly general meeting at the London Tavern, Bishopsgate Street, London. A dividend of 10 percent was declared. They reported that the Cadiangullong Consolidated Copper Mining Company (Limited) had been dissolved and was in the course of being wound up. They further added the property had hitherto failed to be a paying undertaking, but the annexed final report by Captain Holman, gave a fair ground for expecting that this company's interest in it may yet prove to be of adequate value. At the request of J. H. Miller, Manager of the Cadiangullong Copper Mining Company (Limited), Captain Holman wrote a report dated 11 June 1868 which was annexed to the company report. He calculated that copper ores smelted at the works had been almost 7700 tons yielding of 837 tons 11 hundred weight and 6 pounds of refined copper from East Cadia, West Cadia, Canobolas and Carangara mines. He believed the mines' closure was caused by the extraordinary low prices, then prevailing in the English markets for copper as well as inadequate capital to develop the mines on a large scale, and the partial falling off in the quantity and quality of the ores produced latterly from the mines.

In the report, he outlined the property as being over 1570 acres on the east and west side of Cadiangullong Creek, with a manager's house 30 by 60 ft and some 60 huts suitable for officers and workpeople's residences, a few of the latter built of slabs with shingled roofs, the remainder of slabs and bark roofs. There were also two hotels and three stores all the property of the company and a public school standing on an acre of land belonging to the Government. Josiah described the smelting works in some detail:

Smelting Works. – The large shed covered with galvanised iron is 125 by 60 feet, under which are three copper ore furnaces, one roasting furnace and a refinery furnace. Another galvanised iron covered shed 60 by 50 feet, contains one new copper ore smelting furnace complete. A detached galvanised iron shed, 35 by 25 feet, covers a calcining furnace. Detached is a smith's shop with forge, anvil, vice and tools; also sets of smelting tools. An assay office built of slags, with shingled roof 29 by 18 feet, contains two furnaces, assay tools, scales, weights, crucibles, chemicals and fluxes for assaying, with office furniture, stationery, etc. There is an Avery large weighing machine and a weighbridge used for weighing the fuel for the works. These works are capable of reducing over 300 tons of copper ore monthly.

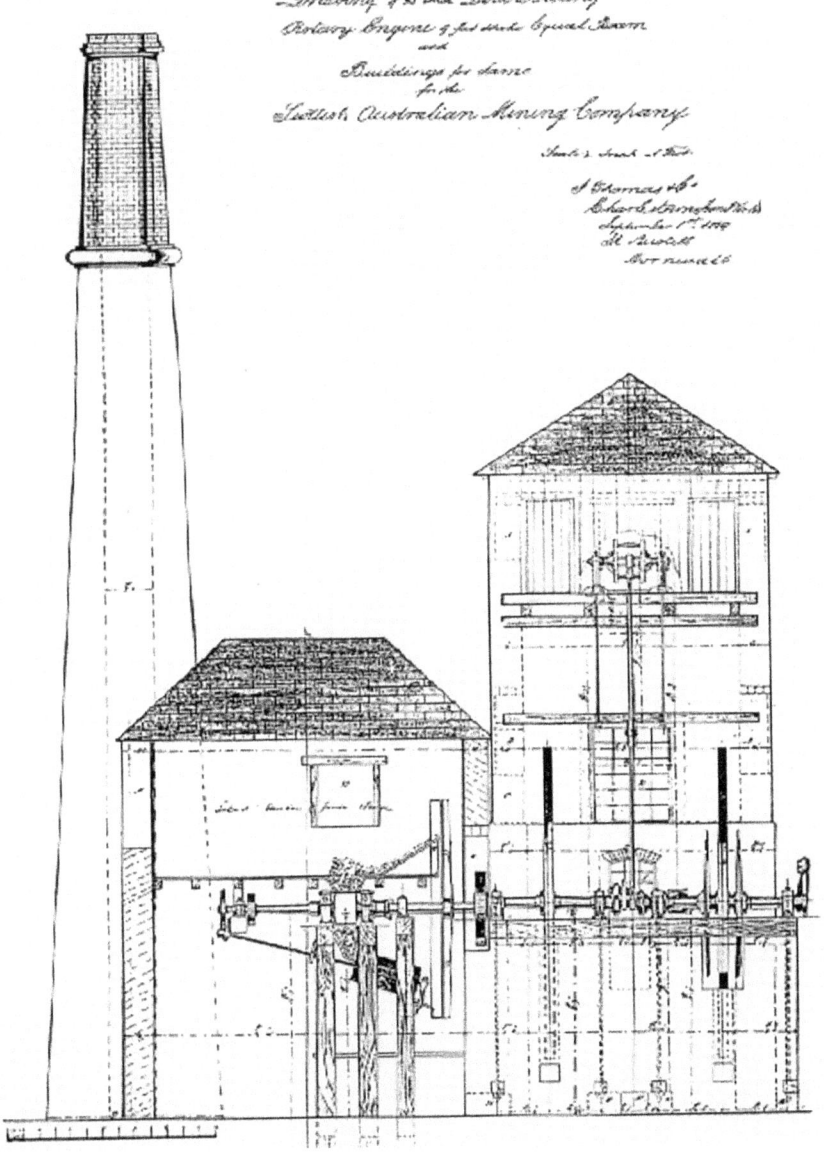

Josiah always had great confidence that the mine had vast untapped wealth and concluded in his report:

> *Both the surface and deeper works here have given ample evidence of the existence of gold throughout this unusually large lode, and, independent of its offering encouraging prospects for giving large returns of copper when developed to greater depths, I have great confidence that the vein, abounding in iron and iron pyrites, traversing the whole length of the property, will ultimately be found to produce gold in payable quantity, and inexhaustible in its supply of auriferous stone.*

At around the same time, gold was discovered at Four Mile Creek just 2 ½ miles west of Cadia. Many of the miners took the opportunity to try their luck as there was very little activity going on at Cadia.

After Cadia

1868–1871

With the mine closed, Josiah explored other opportunities and was asked to take the position of manager of works for the nearby Icely Copper Mine, operating under the name Manton Copper Mining Company, owned by the Western Mining Company. The mine was situated within the Icely Copper mine area near Guyong about 22 miles west of Bathurst. Copper production had increased during the first half of 1868 almost doubling the previous six months, averaging 16 percent. Under the direction of Eynon Deer, smelting operations had been established some years earlier and now four furnaces were in operation which allowed for an increase in the amount of refined copper produced. Josiah's first task was to carry out a complete assessment of operations and report to the directors. On 18 July 1868, he provided a report for the directors of the company, calculating smelting costs to be £2/ 12s/ 1d per ton for ore and £18/ 9s/ 5d per ton for refined copper. Of concern was the amount of water seeping into the mine shafts which would require installation of steam-driven machinery to not only drain the mine but also improve crushing of the stone and ore. This would also allow further development of the mine at greater depths and facilitate the opening of adjacent parallel shafts. Additional shafts had been sunk, the deepest now to 35 fathoms which had recently struck a rich lode. At the time, there were 12 miners engaged at the mine and the yield had improved to 20 percent.

As well, Josiah took on other assignments, reporting on the viability of new mineral discoveries. If his reports were favourable, these were added to company prospectuses in hope of raising capital to develop a mine. He travelled to the Grenfell mine area, 35 miles west of Cowra and visited the famed Emu Creek diggings. These were scattered for many miles where various diggings with numerous stamper machines often shared in a collaboration between various mining groups. Grenfell had sprung up out of the mining frenzy. Shops of every description and public houses formed a guard of sorts, sporadically lining the one main street. Water can halt a mine's progress like a fallen tree can block a carriage. Cadia had too much, Yass had too much, but Grenfell had too little, and at times the diggings were all but deserted. One of the diggings that was progressing well was that of a consortium of Messrs. Everett, Vaughn and others. From Josiah's observation, this seemed the most promising of the mines he visited and after some discussion with the owners, offered to manage the mine should they wish to develop it further.

In September 1868, Josiah travelled south 50 miles from Cowra to Burrowa (now Boorowa) to inspect the Burrowa Copper mine. The directors of the Burrowa Copper Mining Company (Limited), were looking to raise capital to develop the mine further, £60,000, offering 60,000 shares of £1 each. Mining had commenced and ore sold had mostly covered existing costs. There were three main shafts to inspect at 30, 60 and 65 ft. Carefully climbing down the ladder of the first shaft, he noted rich stones of blue and green carbonate of copper ore. The other two shafts, about 100 yards apart also displayed black and yellow sulphates of copper. A well-defined lode could be seen varying in size from 8 to 24 inches when exploring a 48 ft drive into the hillside slope. To date, the lode yielded from one to three tons of copper ore per fathom. Exploring further he noted a number of heaps of ore at grass and estimated this to be around 30 tons. He felt the mine ... *fully warrants an outlay being made to prove it in depth; indications of the lodes, as far as seen, justify the expectation of their proving profitable below.* as stated in his report dated 5 October 1868.

In November 1869, a petition with over 160 of the electors of the town and district of Orange, including Josiah Holman, was signed requested the Honourable (later Sir) Saul Samuel, Colonial Treasurer, to stand as a representative of the electoral District of Orange. He was elected unopposed in December.

In early February 1870, the half-yearly meeting of the shareholders of the Western Mining Company (Limited) was held at Bathurst. Prices for copper had continued to deteriorate and the mine was suffering from continual machinery operational problems. Josiah put forward a number of suggestions to improve the mine's performance, but agreeing to the recommendations would require additional capital. An earlier attempt to raise further capital to purchase heavy machinery to empty the shafts was unsuccessful. He wrote that at the 65-fathom level of Stevens shaft, a fine lode of ore yielding over 20 percent, and averaging from 2 to 3 ft wide, had been cut on the northern side of Stevens shaft. Chairman J. W. Ashworth in his report for the six months to the end of 1869 to shareholders of Western Mining Company, noted what effect the falling price of copper was having on profitability. Ashworth also highlighted the difficulty in raising additional capital required to carry out works to remove water problems identified earlier by Captain Holman.

It was around this time that Josiah was asked to complete an assessment of the quartz reefs for the Enterprise Quartz Mining Company located at Trunkey Creek 35 miles south of Bathurst. At the Enterprise Quartz Mining Company shareholder meeting held on 27 July 1870, directors expressed concern at unsatisfactory results of the past four months of the company. Included in that report was a note as follows: "... the directors, deemed it advisable to request Captain Holman, who is Super intending the Icely Mines, to go over our property and report thereon. This gentleman's report has since come to hand, and is appended hereto."

Josiah always had great faith in the Cadia mine and with very little direction being made by the directors of Western Mining Company, he put forward a proposition to SAMC to work the properties for gold and copper. He was aware SAMC was reluctant to invest further in the Cadia mine, and as gold had been discovered, he proposed to open up the mine to gold-diggers and tributers. They would derive their remuneration from a percentage of the value of alluvial gold, auriferous stone or copper ore which they raised from the reef or lodes. In this way, it was expected the properties would be tested without risking further capital for that purpose. Stamps were readily available on the site to crush the stone and Josiah estimated that extracting half an ounce of gold to the ton of stone, if mined on a larger scale, would be highly profitable. Quartz reefs, in the neighbourhood and up to the boundaries of the Cadia properties, were almost identical in character and composition with the reefs that traverse nearly the whole length of the Cadia properties. These had recently been mined for gold and proved to be very lucrative.

On 26 October 1870, the Scottish Australian Mining Company (Limited), held their half-yearly general meeting for the six months ending on the 30 June 1870. They reported that an opportunity had arisen through Robert Morehead for acquiring, on favourable terms, the company's co-proprietors of the Cadia properties. A conditional arrangement had been made to purchase the property for the sum of £2000. Simultaneously, with the conclusion of this purchase, an arrangement had been made with Captain Holman by which he was to take a lease of the property for 12 months until 30 September, 1871, for the purpose of further developing the metalliferous minerals and lodes therein. He was to pay a royalty to the company of one-tenth part of all copper ores and one twentieth part of all gold obtained from the property. They also added that Captain Holman expects that he will make his operations remunerative to himself and to this company and that he intends to trace the lodes westwards beyond the point up to which they have hitherto been proved to exist.

To the directors Josiah wrote:

In addition to the minor auriferous lode, in point of width or size, I am fully impressed that the iron lode, fully a chain (66 feet) in width, on which copper ores were first worked for in this property, will be proved payable for gold, and should my anticipations here be realised, the quantity of stone easily available is unlimited. Should gold in payable quantity and stone in large quantity be proved to exist in this property during the first year, it will then be necessary to make a large addition to the stamping machinery, and for every additional stamped put up, an additional revenue will accrue to the company.

He added:

In fact, I have little hesitation in stating my opinion that the company will find it to their advantage, at the expiration of my lease, to take over the works on their own account, and to erect a large crushing plant for extracting gold from stone produced from their own property, as well as from stone that will be furnished to crush from adjacent lands, at a highly payable rate of crushing.

On 21 May, it was reported that:

Mr. Josiah Holman, of Cadia, has leased the Cadia Copper and Gold Mining estate, with the engines, plant, and smelting works, and is prepared to let patches for copper and claims to dig for gold.

Returning to Cadia in April 1870, having secured arrangements with SAMC to lease the Cadia property from 9 May 1870 to 30 September 1871, much of his time was spent directing the few miners that remained living in Cadia. Working Holman's Reef on the South end of the mine was on a small scale with only occasional use of the beam engine. His return also coincided with the marriage of his daughter Emily Louisa to William Blood on 20 April in Cadia.

Even though the weather conditions for mining were more favourable in the early seventies, by the mid-1870s the population of Cadia had dwindled to around 150, many of the miners and families having moved on to more lucrative opportunities. There were two churches, the hotel had closed and only one or two shops were still operating, including the post office in Chilcot Street. The school remained open for the few mining families' children and nearby pastoralists. John Christoe, Josiah Holman and William Blood had been appointed as the first members of the School Board. Much like churches at the time, the school not only served the educational needs of the local children, but also acted as a centre for community life. On 24 June 1868, Josiah informed the Council of Education that the mine had ceased operation, but in November 1868 he also organised a petition to keep the school open. Through this petition, the school survived the closure of the mine and served the Cadia community until 1930. The school re-opened briefly in 1943, but was finally closed in June 1945.

As regards gold, it was well-known before the operations of the Cadiangullong Consolidated Copper Mining Company were discontinued, that the property was more or less auriferous. Some quality rich specimens of auriferous stone had been found scattered over various areas of the surface, but the vein from which they were perhaps to have originated had never been discovered. In the hope of searching deeper, Josiah proposed SMAC invest £500 for heavier machinery. This request was rejected by the board.

In mid-August 1871, with the lease arrangement with SAMC now almost expired, and not wishing to renew it, Josiah accepted a request asking if he could examine a leased 45 acre site near Scone, NSW. During the week of 21 August 1871, he travelled from Cadia to Bathurst and then 35 miles further on to Rydal to catch a train to Sydney where he boarded the overnight coastal steamer to Newcastle. From Newcastle, he caught the 7 a.m. train to Scone. arriving just after midday. Finding a livery nearby, he secured a horse and after purchasing some fresh provisions, left for the 35 mile journey to the tiny village of Moonan Brook. He had been engaged to examine two of the 45 acres, 10 acres each, of the leased gold-mining properties known as Fuller's Reef, part of the Denison gold mining area. The quartz reefs had been profitably worked for a number of years, the stone averaging at times as high as nine ounces of gold to the ton. Construction work had been undertaken to establish the layout to facilitate the delivery of large quantities of stone. A tunnel, 522 ft in length had already been completed, with plans for it to be extended to 840 ft until it reached the "whim shaft" at the other end of the lease where a tramway would then be laid. The reef in places was 3 ft thick, but sometimes dwindling down to just a few inches. The company, Denison Gold Mining Company, yet to be listed, was looking to raise £30,000 in 30,000 shares of £1 each to purchase the property and replace the present plant with new, more efficient machinery. From Scone, Josiah rode east, crossing many creeks and rivers on a road hardly capable of supporting a dray, more like a bush track in places. Added to that, he had to stop continually at the many gates of different and varying contraptions, offering challenges for the uninitiated to open. As darkness replaced light in the rolling valleys and not knowing the road or terrain, he decided to camp on a river flat beside a small creek, one of many. Continuing on at first light, he passed Belltrees, stopping momentarily to admire the substantial homestead of Belltrees Station. He was pleased he had decided to make camp the previous evening well before dark as the road continued to deteriorate, the countryside more undulating, before he reached Stewart's Brook and then Moonan Brook.

At Moonan Brook, Josiah observed a number of wood slab huts scattered about and nearer the creek shaded by heavy timber, a hotel, the Reefs Arms with a store and post office. A crushing machine of eight heads, showing the wear of eight years devoted service, stood neglected next to a boiler, its time well passed, as did the 16 ft crumbling water wheel nearby. Further along the meandering creek, the start of a new public house, Willow Tree Inn owned by Daniel Cook, was in the midst of construction. Travelling about two miles, on the South Lease of the mine section, he inspected an adit running northerly along the course of Fuller's Reef underneath the old workings. It ran for about 440 ft, attaining a depth of about 250 ft at its deepest, the reef itself varying in size from six to thirty inches, displayed some visible coarse gold seams running through the solid white quartz. Walking further to the sloping North Lease, he passed the old workings, now filled with water, and determined the adit would need to be extended a further 300 to 400 ft to allow adequate drainage. This would also present an opportunity to extract a significant quantity of auriferous quartz. Talking with a number of the miners nearby, Josiah learnt that the North and South Leases were separated by a claim belonging to Messrs. Warren and others. He watched as the miners crushed 21 tons of quartz which yielded 5 oz of gold to the ton, quite a good result. After some further exploration of Warren's lease, he determined there would be a number of benefits in purchasing the lease, the most obvious being extending the adit crossing through the lease.

Having completed his review of the Denison mine site, he retraced his journey back to Scone, returned the horse to the livery and found lodgings at the Belmore Hotel opposite the railway station. Rising early, he caught the 4.50 train to Newcastle, arriving at 12.15 in the early afternoon and walked to the nearby wharf to catch the overnight steamer to Sydney. Having completed his report while travelling, he visited Denison's Agent at 131 Pitt Street and left the report with Mr Joseph Mullens.

> **THE DENISON GOLD-MINING COMPANY (LIMITED).**
>
> Denison Gold-Fields, thirty miles from Scone.
> To be registered under the Limited Liabilities Act.
>
> Capital, £30,000; in 30,000 Shares of £1 each, 10s paid.
>
> Provisional Directors:
> John Levien, Esq.
> S. A. Joseph, Esq.
> Oliver Saunders, Esq., Jerry's Plains.
> Mining Manager:
> F. A. Martyn, Esq.
> Legal Manager:
> J. Oswald Gilchrist, Esq.
> Bankers:
> Oriental Banking Company.
> Solicitors:
> Messrs. Roxburgh, Slade, and Spain.
> Broker:
> Mr. Josiah Mullens, 131, Pitt-street.
>
> This Company is formed for the purpose of purchasing and working the well-known Fuller's Reef and other quartz reefs on the Denison Gold-fields, which have been profitably worked for some years past; together with the present plant, buildings, machinery, and freehold and leasehold lands belonging to the proprietors, say about forty-five acres, as detailed in the accompanying schedule.
> The report and plan annexed will show that an immense amount of work has already been done towards laying out the mine for delivery of large quantities of stone, and the results already obtained show that it is likely to be second to none of its kind in New South Wales in its dividend-paying prospects.
>
> Extracts from Mr. Josiah Holman's Report upon Messrs. Levien and Saunders's Claims, Moonan Brook, Denison Diggings, New South Wales.
>
> "Sydney, 4th September, 1871.
> "During last week I duly examined the two 10 acres leased gold mining lands belonging to the above firm. There are situated about two miles from the township of Moonan, in a mountainous district, that offers great facilities for the economical extraction of the auriferous quartz by means of adits driven on the course of the well-known Fuller's Reef, which has been proved to traverse the whole length of these claims.

Sydney Morning Herald 10 March 1872

After lunch, Josiah walked to Central Station and caught the 5 p.m. train to Goulburn arriving at 11.30 where he stayed overnight. In the morning, he hired a horse and rode some 25 miles south along Braidwood Road, through the small village of Tirranna until late afternoon when he reached the Currawang mine site, near Lake George. Copper had been mined here, on and off, since 1865, but more recently had been mined *with more vigour*. Over 1000 tons of ore had been extracted with an average yield of 14 percent, not considered payable quantities at current low prices. Earlier yields exhibited copper yields of almost 28 percent with lead sometimes 11 percent and silver 9 oz 16 dwts [Pennyweights] per ton of ore. A smelting furnace had been constructed in late 1867, but found to be wanting and the following year was altered and an additional six new furnaces were erected by Eynon Deer, the chief smelter. Expectations were not realised by this additional expenditure and Deer was blamed, the owners refusing to pay. Having lost the case against Deer, the company wound up and Eynon Deer was awarded leasing rights which were to expire on 30 April 1872.

J. B. Watt and John Frazer, directors of Phoenix Copper Mining Company, previously directors of Canobolas mine, had contacted Josiah and asked him to provide an assessment and prepare a report on the mining operation and its likely prospects. The directors had formed a company, the Phoenix Copper Mining Company Limited and had come to an arrangement with Eynon Deer to acquire the Currawang site. They now wished to offer 50,000 shares to the public to raise capital to purchase equipment to further develop the mine. Josiah arrived mid-morning of 5 September and met with Deer. Eynon Deer was well known to him. They had worked together previously when Josiah managed Icely Copper Mine and used the Cooperative Copper Smelting Company under the management of Deer, then the chief smelter.

Once Josiah found Deer at the mine site, he was greeted warmly and both continued chatting as they headed towards the manager's cottage. Josiah could have stayed at the nearby Telegraph Inn, but Derr insisted he stay in the cottage with him and Mrs. Deer. They spent the rest of the day inspecting the smelting works, furnaces, engine room and blacksmiths furnace. Josiah examined the various shafts and adits that had been quarried by the 80 or so men recently engaged by the company. Excavations were shallow, exposing the extensive copper lode from which stoping had been carried out at the 10 fathom level for almost 160 ft. A considerable amount of ore was stacked in heaps nearby amounting to around 700 tons, Josiah estimated. On the western side, a shaft had begun which exposed a copper lode about 13 ft wide running to almost 33 ft. Josiah was particularly impressed with the purity of the ore, and estimated a further thousand tons of ore could be easily stoped down to the 15 fathom level. He continued taking notes as he and Deer wandered around the site examining various aspects of development.

The following day, Josiah examined the 16 hp engine and hauling gear, ensuring it was in suitable working order and capable of handling excavation to at least a 20 fathom depth. He noted an additional pair of rollers for crushing the copper ore should be procured. He also spent some time with the blacksmith and both agreed the furnace would need extending to handle any increased activity. After completing his survey, he determined the smelting works, consisting of a large calciner, three ore melting furnaces, two roasters and refinery were in good working order. All were covered with a galvanised iron roof which extended over the shed. In his opinion, the works were capable of treating about 300 tons of ore per month. While inspecting the smelting works, he was advised that 15 tons of coarse copper and regulus were in various stages of treatment.

In the afternoon, he inspected the books noting that over the past year, almost 4000 tons of ore had been extracted and smelted, producing 290 tons of refined copper with an average yield of 12½ percent. He spent the rest of the afternoon riding further outside the property and was pleased there seemed to be plenty of timber for the furnaces, of which almost 5000 tons had been already stacked nearby. Securing carriers to collect and transport wood for the furnaces was becoming more and more difficult within the mining industry. The following morning, having completed his assessment, he thanked the Deers for their hospitality and returned to Goulburn. After returning the horse he caught the mail coach to Bathurst and stayed the night at the Black Horse Inn in George Street. At Bathurst, he called in at the Cobb & Co. head office and determined he would have enough time to return to Cadia for a few days, wash and collect new clothes, catch up with family and any news before heading off on his next assignment.

With the closure of the mine many of the residents of Cadia village had moved on, some embarking on different careers within the Orange area. Others readily found employment in the numerous mines being discovered locally, such as Carcoar and Hill End and further west as far as Cobar.

On May 8 1874, at the half-yearly general meeting of the Scottish Australian Mining Company, it was agreed to spend £2000 on further exploration at Cadia, as some gold had been found there by the men who had been developing a small part of the property. Over the 30 years from 1869 to 1899 the village survived, mainly supporting agriculturists and the occasional prospector and small-scale gold miners. The village was sustained as much, if not more so, by the growing agricultural activity in the area. However, at no stage did it approach the scale of the 1860s.

Josiah Jnr. was sent to Newington Boarding School in Sydney aged 15. He started on 28 July 1868, admission number 174. It can be assumed that Charles William continued school in Cadia, there is no record of him attending any other school. Annie attended the Cadia school before being sent to boarding school in Petersham in 1871.

New England

1871-1873

Josiah had arranged to meet William Corbett Lawson in Tenterfield around 20 September 1871. Corbett Lawson, a magistrate, landowner and entrepreneur, had been involved in various mining ventures and lease holdings around NSW and was representing a number of Sydney investors, including Saul Samuel, John Frazer and Simon Zollner. Tin of high quality had reportedly been discovered in the Tenterfield district but the investors wanted Josiah's opinion, hoping that if he judged the area a sound investment he might act on their behalf. They knew of his tin-mining experience in Malaya; tin after all was in his blood. Much of his earlier mining experiences had been working the tin mines around Gwennap Cornwall, then considered the richest in the world, hence he was very keen to visit the area and determine the validity of the exceptional claims being made. Tin had been found as early as the mid-1860s around Oban, as well as Inverell and Tenterfield. A number of determined prospectors had been barely scraping by on a meagre existence ever since, including Chinese miners who had moved from earlier diggings around Uralla. Gold had also been discovered as early as 1866 around Cameron's Creek, a rugged hilly area which could only be reached on foot. In fact, more than 100 ounces of gold had been brought to Armidale in nuggets of varying sizes in recent years. But mining on a commercial basis had not yet begun.

Fog hung low over Cadia Valley as Josiah waited in the early morning for the mail coach to Orange. It would be a full day's journey to Bathurst and from there to Rydal where he would catch the train to Sydney, staying overnight. The next morning, he boarded a steamer to Newcastle and then a train to Scone where he caught the Cobb and Co coach to Tenterfield via Tamworth. Corbett Lawson meanwhile had caught a steamer to Richmond Heads (Richmond Heads), leaving Sydney on 8 September and arriving in Ballina on the evening of the 10th. From there, he travelled to Tenterfield via Lismore and spent some time talking with miners and visiting a number of the claims in the area. On Friday 22 September, Josiah and Corbett Lawson hired a horse and buggy from the livery in Tenterfield and headed east 30 miles to the small village of Drake where they examined a number of reefs that had already been excavated. Both were suitably impressed by the excavations and shafts that had been constructed.

They returned on Saturday to Tenterfield, where more than 40 miners had gathered to meet Corbett Lawson at the Perseverance Hotel. He had provided a crushing machine for the miners earlier in the year, but it had been plagued with mishaps and defects and there had been considerable delay in getting an engineer to Tenterfield for the necessary repairs. Corbett Lawson had engaged a Mr Muirson to examine the machine and he had determined what alterations were required to ensure the stampers were in good working order. The miners were frustrated by the continual delay. Without crushing machinery, they were faced with considerable expense, £7 per ton, to cart the ore to Grafton — a three-week treacherous journey there and back. Mr Horton chaired the meeting and Corbett Lawson addressed the agitated men, promising a first-class crushing machine would be provided that had the capability of crushing one ton of ore an hour. He further guaranteed the miners he would not charge more than 30 shillings a ton to crush any excavated stone and rock. He also promised, if the miners agreed to his terms, he would have the machine operational by 10 January 1872. Mr Saunders proposed that the miners accept these terms which was seconded by Mr R Laird and unanimously carried. The meeting concluded with the miners signing an agreement with Corbett Lawson.

Early the following morning, Corbett Lawson and Josiah caught a coach to Inverell where, the following day, they inspected a tin mine in the area before returning to the New England Hotel. Claims had been made on most of the area where alluvial tin was being mined. Josiah spent some time talking with a local miner, Mr C. S. McGlew on his selection near Elsmore, 10 miles from Inverell. A 40 ft shaft had been excavated to test the depth of the lode. Josiah was impressed and determined the almost black alluvial tin shoal could be easily mined in quantity with the right application of machinery. Already, McGlew and his men had begun to dam creek tributaries to set up a sluicing-type operation to wash the ore for bagging and Captain Barron had ordered machinery; a boiler for a steam pump and a large quantity of piping expected any day soon. Josiah spent most of the next day examining other creeks, mainly Copes and Darbys around the Tingha area where he met licenced surveyor Mr J. M. Simpson, who, with a small team, was measuring some selections. A number of miners had already established a camp and Josiah was more impressed by what he saw in this area, and given very few claims had been made, set about mapping out areas to register a claim. Having completed his exploration of the creeks and gullies of interest, the next day he and Corbett Lawson travelled by Cobb & Co coach back to Glen Innes, where they stayed the night. The following morning, they continued on to Grafton, boarding the 110-ton *Helen McGregor* steamship on 9 October arriving in Sydney on 11 October.

In Sydney, Corbett Lawson quickly contacted his associates and arranged a meeting with Josiah to discuss the findings of their visit to Inverell. Saul Samuel, John Frazer were immediately interested and contacted others, J. Brewster, F. H. Dangar and C. A. Fraser. All agreed to fund the project and establish the Newstead Tin Mining Association. After securing a lease of over 1000 acres, Mr W. L. Jenkin was approached to manage the assaying and Mr J. M. Simpson the mining works. On 14 November, it was reported that the Newstead Tin Mine had discovered a rich lode around the Inverell district which created much excitement amongst investors. It would take until 23 February 1872 before a company was listed and a prospectus published:

NEWSTEAD TIN MINING COMPANY (Limited)

Inverell, New England District.

CAPITAL, £60,000, in 60,000 Shares of £1 each. Issued at Ten Shillings paid-up.

DIRECTORS. John Brewster F. H. Dangar, John Frazer, S. Samuel, C. A. Fraser.

Local Director. Duncan Anderson. Mining Manager W. L. Jenkin.

In compliance with a strongly expressed desire to obtain shares in this Company, the proprietors have consented to place 10,000 shares on the market at their par value of ten shillings paid-up, on the following terms.

Deposit, 2s 6d on application, 2s 6d on allotment, balance 6s on signing Deed of Settlement.

The property of the company consists of about 160 acres of land, situated near Inverell, between the McIntyre River and King's Creek, and in the immediate vicinity of the Elsmore mine. Nearly the whole area of this fund has been found to contain alluvial tin, together with stanniferous dykes or lodes yielding rich black oxide of tin.

There is now at the credit of the company £4000 of subscribed. capital, after paying all outlay for securing and sinking shafts, and other works on the mine. This, it is deemed will be amply sufficient working capital to put the mine in the position of a dividend-paying investments.

During the meeting in Sydney on their return in mid-October, Josiah also related conversations he had with a number of the miners while in Tenterfield who had spoken of tin discoveries around the small village of Oban, some 90 miles south. He was told claims that had been made on the most likely areas in and around Paddy's Gully and Mount Mitchell and it seemed a number of leases had already changed hands. It had been reported that George Taylor and Company had sold a gold lease in the area for £100 to the Sydney Tin Mining Company. Already, Josiah was told, an Inn and a wooden slab shop had been erected by John Clark to sell food and goods to the miners and John Wheeler had opened up a hotel, Digger's Home, at Paddy's Gully. Tin ore, in a streaming fashion was easy to work the miners had said, two men working together could quarry a ton of ore in a week. There was also the bonus of finding a small gold nugget or two. The men at the meeting were immediately interested and asked how soon Josiah could return. Terms were agreed, Captain Holman was to act on their behalf and using his considerable tin mining experience, make claims on the areas he considered worthwhile. The next day they met at the solicitors' offices in Pitt Street and signed a Deed of Agreement granting him permission to act on their behalf.

When Josiah arrived at Oban, he discovered that indeed a number of claims had already been made. After exploring the area's escarpments, depressions and valleys, studying the drainage and following the ridges and hollowed-out granite, he was able to assess the most promising areas. In the most likely prospecting areas where no claims had yet been made — around Paddy's Gully, including Backwater and Sara tributaries — he pegged out claims and within a few weeks, had acquired over 3000 acres of leases on behalf of the Sydney investors. Some of the more promising ones he purchased were £200 for Brunnell's and Bacheldor's lease and £300 for Heghan's. Having secured the leases, he quickly engaged a number of miners willing to work on a tut or tribute basis and directed them to the allocated site areas where they were to work. By November, the miners he had engaged, had excavated more than 10 tons of ore sent by dray to Sydney for crushing and smelting. So confident was Josiah of the prospects of the mining area, and the exceptionally high quality of the ore, he proposed and gained agreement from the Sydney investors to develop a smelting operation.

Other mining companies had been established to explore and work the Oban area. They included Messrs Bagot and Company (which had discovered a rich lode just 4 ft below the surface at the head of Snakes Gully) Clark and Company and John Moore and Company of Armidale which held large leases on Horseshoe Bend. Thomas Phillip Clark, licensee of the Golden Age Hotel, had earlier staked a claim on over 400 yards along the main creek at Oban. Other successful leases such as Bagot and Co, a Glen Innes company and Starr and Co, managed by Captain Edwards, were having considerable success sluicing stream tin, with good future prospects based on sample of heavy ore. Another, Markham and

Co, also sluicing, was showing very good prospects as were a number of smaller claims owned by local Armidale business people and miners. On 2 December, two auriferous leases adjoining Josiah's Mount Mitchell leases were offered for sale by W. C. Rodgerson, of Glen Innes. The area around Oban now supported more than 600 miners and prospectors, with a growing Chinese population, many having moved from the Uralla workings. Demand for mining equipment and other goods exceeded supply and Josiah saw an opportunity to open a store next to the home he was building. Materials for the building were ordered, with prospecting tools for the store being procured from Armidale and Sydney.

On 17 November, Mr John Buchanan, Gold Commissioner for the Northern Goldfields responsible for the granting of mining leases visited Armidale and met a number of disgruntled miners. The following day he travelled to Oban and spoke with the miners, discussing their grievances regarding mining leases. Mr T Wilson spoke on behalf of the miners, the main complaint being large Sydney corporations taking up huge areas leaving nothing for the struggling miners. On 25 November 1871, *The Sydney Mail and New South Wales Advertiser* wrote:

> The greatest drawback is the want of information of what lands have already been applied for, and as long as the applications have to be made at the office of the Minister for Lands in Sydney much ignorance will exist among the searchers, and many disappointments will be the result, as it can only be after the Surveyor's chain shall have passed over the ground that the vacant portions will be known.
>
> Many and handsome are the specimens that have already been exhibited about the town. It is a most fortunate circumstance that each possessor of a claim or lease is sanguine of making a fortune thereby — but this is certain, that the ground is very rich in ore, and very valuable discoveries have already been made. And we may expect even more, as, where such large quantities of the metal are to be found, as at Paddy's Gully and elsewhere, a vein or lode cannot be far off.

On Friday last week a deputation of miners waited upon the Gold Commissioner, and on the following day a large gathering took place at Paddy's Gully, where the grievances of the gold miners were represented to the Commissioner by Mr T. Wilson, who spoke for some time, the chief complaint being that the Government had allowed mineral leaseholders to take up their claims, and thus possess themselves of large areas of the richest gold producing country at Oban, while the miners were confined to very small areas and under stringent laws. Mr Wilson asked the Commissioner if he would use his influence to that protection, and a discontinuance of the leasing system, otherwise Oban as a gold field would be destroyed.

By reply, Mr Buchanan urged the miners to a strict observance of the laws as regarded their claims and water rights, the latter of which he considered as great privileges which holders of leases could not touch, and urged registration of the same at once. In reply to a question touching the right of the miners to the tin in their claims. Mr Buchanan expressed it as his opinion that a digger's claim was an indefeasible right against all comers, and he believed that no one could touch any material within it excepting the holder. However, he would lay these things before the Government, and, on receiving replies to the several matters that vexed the miners, would communicate to them at once.

His comments did little to appease the miners, and, with lack of clarity over leases, conflict boiled over around Oban. Frequent visits by mining Inspectors helped to determine miner's rights and resolve lease boundaries and disputes. Unfortunately, it soon became apparent this was not enough and Inspector Brown was compelled to appointed Senior Constable McDonald to keep the peace.

The vagaries of climate affect farmers just as much as miners and a drought towards the end of 1871 began to impacted the sluicing operation of streaming tin. In an attempt to overcome this, Bagot and Clark, two of the larger leaseholders, applied to have a water race constructed from Ben

Lomond Lagoon, a distance of over 12 miles. This of course did not halt progress, already Bagot and Clark had sent two tons of ore to Sydney using blasting where the hard rock could not be mined with hand tools and sluicing was not possible. The crushing plant, now in operation sent over 50 tons of high-quality ore to Sydney realising from 50 to 68 per cent once processed, according to assays made by Professor Watt that would fetch from £70 to £80 per ton in England. As well, there were 12 to 14 tons of ore ready to be carted to Sydney, the ruling price of tin between £50 to £80 per ton. Josiah was so impressed, that when offered the position as mine manager, he negotiated with Mr Zollner, director, to be paid a reduced rate compensated by shares in the company when listed.

As development of Josiah's mining site was moving rapidly, too rapidly for some, a number of concerned investors asked director Simon Zollner to visit the area to ensure their investment was in good hands. Already a large shipment of heavy mining equipment had recently been sent from Sydney and crushing operation had begun. Zollner was impressed by what he saw, but not so the extremely rough road over rugged terrain through thick wooded hills from Armidale. Closer to the mining activity, he passed the beautiful cattle station homestead built by Robert Rummage for Andrew Coventry who had previously worked for many years as superintendent at Saumarez homestead.

By the end of 1871, the area had developed into a hive of activity, hundreds of miners vigorously working their leases. A post office had been in operation since 1869 and construction was almost completed on Mr P. Mahony's Inn at Paddy's Gully and Mr Fi Long, a Chinaman was already operating a general store. At Backwater, Messrs Starr and Morrow were erecting a number of buildings which would become an Inn, residence and butcher shop alongside John Clark's new Pick and Shovel Hotel. Josiah had engaged workmen to build his own residence along Paddy's Gully as well as a mining equipment store which was soon trading well. Such was his belief that this mining area would continue to be productive for many years to come. Near Uralla, 12 miles south of Armidale, Chinese goldminers had been very active when the discovery of alluvial deposits, over a decade earlier, had seen the population grow to more than 3000. Miners, disappointed with other discoveries, flocked to the

Uralla hoping this would be the new El Dorado. A correspondent writing for the *Sydney Morning Herald* when visiting Oban at the end of June 1872 wrote; "Strangers visiting Oban now would almost fancy that they had got into some part of the Chinese territory. The Celestials lately have become very numerous, for the reason that they work for less wages than a European, and, from all I can see, do just as much work."

Painting by William Gardner of Oban Station located between Bald Blair and Mt Mitchell.

Returning to Sydney in late December, Zollner reported considerable progress had been made under Josiah's management. He agreed with him that the mine was indeed a very rich one, and progress should continue without delay. Most of the ore had been purchased through tut and tribute agreements with a number of the miners with selected areas being offered on tribute at an agreed price of £18 per ton. The mineral laden ore was packed in canvas bags each holding about 100 lbs each, then loaded onto bullock drays for the journey to smelters in and around Sydney. Each dray could carry about 2 tons over the most difficult terrain to either the ports of Grafton or Morpeth. By the end of the 1871, Josiah managed the largest number of leases and was employing over 20 men, prospecting and washing ore. As is the way in this country of contrasts, heavy rain at the end of January 1872 ended the drought and saw the dams fill and the streams spill which again halted mining progress.

In February 1872, three drays left Oban with over seven tons of ore for the port at Morpeth at a cost of £6 per ton cartage. A further six tons were being loaded at Mitchell River with a number of mineral heaps at grass, around 20 tons, ready for dispatch. The *Newcastle Chronicle* reported that on Monday 19 February, one ton of tin ore from Oban that was shipped from Morpeth to Sydney and smelted at the Burwood works produced six large blocks of, "apparently, most excellent metal". With so much ore being mined in NSW, a new smelting furnace was constructed in Sussex Street Sydney, by Milne Brothers capable of smelting one ton of ore per day.

Simon Zollner, having returned to Sydney and enthused by what he had seen, called a meeting of the investors and recommended they float the company. It was agreed a new company, the Mount Mitchell Tin Mining Company, be incorporated and directors, E. Vickory, G. W. Allen (MLA), John Keep, Alex Stuart and Simon Zöllner be appointed. On the 24 February 1872 a notice was printed in the *Sydney Morning Herald* outlining the prospectus for the Mount Mitchell Tin Mining Company Limited. The property was described as 2400 acres, leased from the Crown, situated on the banks of the Mitchell River and its tributaries, including Paddy's Gully. This afforded ample supply of water for sluicing without the necessity of erecting large and costly pumping machinery. Water rights had been purchased, and Captain Holman had erected machinery for the crushing of ore, of which 40 tons had already been extracted. The assay report by Professor Charles Watt, Government Analyst, was included and already one ton of exceedingly fine quality tin had been sold to Messrs P. N. Russell at £158 (equivalent to about A$40,000 today). Tin ore was being raised daily with very little additional outlay. The whole 2400 acre claim is believed to be "richly impregnated with tin ore, and will take many years to exhaust." the prospectus claimed.

Capital was to be raised via 75,000 shares of £1 each, of which 50,000 shares, fully paid up, were allotted to the proprietors, and the balance, 25,000 shares offered to the public on the following terms: 2s 6d per share on application, and the remainder (if required) in calls not exceeding 2s 6d per share, at intervals of not less than three months between each call. Added to the notice was a note:

Captain Holman, a competent and professional tin-mining manager, after inspection of the property, was so thoroughly satisfied, that he stipulated for a reduced salary and a share in the profits of the Company.

By 9 o'clock on the 24th all allocated shares had been taken up, brokers reporting shares later in the day were sold at a premium.

But all was not what it seemed in the burgeoning business colony. Throughout the country, mining discoveries heralded new finds of varying richness, often extremely exaggerated which only fuelled further mounting excitement and anticipation. In the city, companies rushed at a frenetic pace to list in a bid to raise much needed capital. John Buchanan, Gold Commissioner for the Northern Goldfields, was having great difficulty keeping up with the registration of claims, as did those preparing prospectuses, making sure they met the stringent requirements of Company Law. In a haste to *go to market*, there was no time for full and complete due diligence and short cuts were frequently taken. Such was the case with the Mount Mitchell Tin Mining Company when a meeting was called for Thursday 21 March 1872 to approve appointment of directors and agree on the terms of the Deed of Settlement, etc. as required. Disagreement over shareholder dividend entitlements ensued in which legal positions from both sides were argued robustly, but no agreement could be reached. It was finally put to the meeting, and carried unanimously that Messrs S. A. Joseph, W. Gibbs, and F. J. Thomas be appointed as a committee to confer with the original promotors of the company, with a view to the settlement of the existing differences between the two classes of shareholders.

In early March, Josiah received a letter from David Marks, provisional secretary of the Great Northern Tin Mining Company, Horseshoe Bend, Mitchell River, Oban, asking if he would be willing to accompany Mr George Taylor to inspect and report on the mining leases of the company. He immediately agreed, being familiar with the site, as it operated next to the Mount Mitchell mining company's operation he managed. Marks replied thanking him and outlining what was required. Terms were negotiated and on receipt of a letter of instruction and payment received from David Marks, Pitt Street Sydney dated 15 March, Josiah met Taylor and inspected the 280 acre mining site.

Having completed the inspection, he wrote his report on 18 March and posted it to Marks with a copy to John Moore of Armidale, one of the directors. On 23 March 1872, a notice was printed in the *Empire* newspaper outlining the prospectus for the Great Northern Tin Mining Company, Horseshoe Bend, Mitchell River, Oban (Limited). The property consisted of 14 blocks, each 20 acres of land leased from the government situated along the Mitchell River around the area of Horseshoe Bend and Long Point, Oban. These were the leases owned by Messrs John Moore, O. T. Bagot, T. Clark and Co., representing Armidale business people. Capital was to be raised by issuing 30,000 shares of £1 each, of which 20,000 shares (paid-up) were to be reserved for the proprietors, and £1000 each to be paid to them for the purchase of the lease and preliminary expenses. The remaining 10,000 shares were offered to the public on the following terms: a deposit of 2s 6d per share on application, and 2s 6d per share on allotment, with further calls in sums not exceeding 2s 6d per share (if required) in periods of not less than two months.

The prospectus further proclaimed that the area had been profitably worked, realising up to 7 lb of tin ore to the dish in which gold had also been found in every instance. The position of the mine was unequalled, it further stated, being situated at the bend of the Mitchell River whose downward course cut through the Mount Mitchell Tin Mining Company's property and the Oban Tin Mine property. It had the greatest advantage as the natural fall for ground sluicing purposes was not procurable in any other section of the Mitchell River. Below the site was the first block of land purchased by the Mount Mitchell Tin Mining Company, out of which so much tin ore and gold had already been obtained. The proprietors were of the opinion that, in three months from commencement of operations, sufficient tin ore and gold would be raised to cover all the outlay that may have been entered into. All that would be necessary to begin work, was to cut a race at a cost of £200 to £300 and alter the present course of the river, which from its serpentine position through the property would justify. This may well lay open for operations one of the richest deposits of stream tin yet discovered in the colony, together with considerable quantities of gold, the value of which the subjoined report of the Deputy Master of the Mint would testify the prospectus proclaimed.

Royal Mint Sydney, '18th March, 1872.

Sir–The specimen sent by you to be assayed for Gold and Tin was found to contain Gold, 1 oz 8dwt 18gr per ton. Tin, 71.48 per cent.

I am, Sir, your obedient servant,

C. ELOUIS.

Captain Holman's report read as follows and accompanied the Prospectus: -

To David Marks, Esq, 159 and 161, Pitt-Street.

Captain Holman's Report, Mitchell River, Oban, 18th March, 1872.

"Dear Sir–

In compliance with your letter of the 15th instant, I accompanied Mr. George Taylor and inspected the mineral leases held by yourself and company, comprising fourteen 20 acre blocks at the Horseshoe Bend and Long Point, situated on either side of the Mitchell River, Oban, and have pleasure in briefly reporting upon its prospective value for Tin Ores and Gold.

The Mitchell River for over two miles immediately about these blocks has recently been proved to traverse rich Tin bearing country, and the Horseshoe Bend is the first plateau of similar extent found for the reception of its alluvial deposits: and these exist more or less on all the flats throughout these leases which includes the river bed, also for about a mile and a half long, exclusive of one 20 acre block intervening.

The general character of the district is mountainous, the formation true granite rocks traversed by eleven dykes and quartzose veins, thus from Mount Mitchell situated 3 miles up the river, thence to the Horseshoe Bend, the Mitchell and its tributaries traverse gullies, ravines, and gorges which from the disintegration of the rocks have deposited quantities of Tin ores and gold in the flats. The Mitchell ranges continue down to and below the Crown Mountain, the latter forming your northern boundary, whilst on the south gentle slopes give easy access to it. While developing the alluvial deposits which is the primary object, it is reasonable to hope that the permanent sources of tin bearing rocks or lodes will be discovered. No unusual or expensive mining works are required before early and continuous returns of tin ores and gold can be relied upon, beyond cutting short head races for diverting the stream from its bed when it can be applied to useful hydraulic purposes, and in blasting short bars of granite here and there in the river bed as the work proceeds, in order to drain and get at the rich deposits in the recesses of the river bed and alluvial flats adjacent.

I saw prospects of tin ores and gold washed out of your blocks of first-class quality, and I consider its general value promises to equal any ground of similar extent on this river or in the district of Oban. By a moderate expenditure of capital applied by judicious and energetic management in developing these lands for tin ore and gold, I have every reason to believe that highly profitable results will be obtained within a few months.

I am, Sir, yours obediently,

JOSIAH HOLMAN,

Mining Manager to the Mount Mitchell Tin Mining Company, Oban."

Under Josiah's direction, mining continued at a rapid pace on the Mount Mitchell Company property, the tributers quarrying and sluicing between 15 to 20 tons of ore each week. Josiah continued prospecting various areas of the 2400 acre lease, allocating areas the tributers could mine. In April, 25 tons of tin ore were sent to London via the 1094-ton *Christina Thompson* with a further 12 tons awaiting shipment. In transit via dray to Grafton and then to Sydney by ship was a further 17 tons of tin ore; it seemed each day a heavily laden dray left the site for either Grafton or to the port at Morpeth or the rail station at Murrurundi. To help with the busy mine and store, Josiah asked

William Blood (his son-in-law married to daughter Emily Louise in 1870) to join him as mine and store manager in his absence. William owned a once-thriving store in Cadia before the Cadiangullong mine was closed. To Josiah's delight they agreed; there was very little activity going on in Cadia then. The family moved into the residence at Paddy's Gully with their young daughter Mary Ann. Josiah's hard work was paying off, and with the exceptional price of quality tin, Mount Mitchell Tin Mining Company was able to declare an interim dividend of five per cent at the end of June 1872.

At the same time, the Mitchell River Tin Mining Company wishing to be listed and, having received reports from Josiah and Captain Edwards, posted the following notice in the *Sydney Morning Herald*:

"These lands will favourably compare in production of tin and gold with the adjacent mineral leases, one of which, known as Taylor's gold claim, and which were worked solely for its gold production for upwards of a year prior to its purchase by the Mount Mitchell Tin Mining Company, has since it became the property of the latter Company, produced upwards of ten tons of rich tin ore, at a cost not exceeding £18 per ton, obtained from the northern side of the stream, without even diverting the water, or even touching any tin ores from the river bed. These selections can be profitably worked with 20 men, and £460 will purchase the necessary machinery. Immediately after the plant and staff are at work, profitable results should follow."

Water channels from the dams or creeks were redirected by blasting and tunnelling or where the topography did not allow the construction of channels, wood was used to create the races. The channels ran into wooden boxes for sluicing, the water washing away the topsoil and clay, the remaining non-ferrous gravel and stone (tailings) then cast to one side. Josiah was very familiar with this type of operation, having visited Malaya some 15 years earlier when assaying tin for Bolitho of Penzance, Cornwall and Enthoren of London. One evening when returning home, he stopped by the lease operated by Chatfield and others and suggested re-washing the tailings, as the price of tin had recently increased significantly from around £18 to £45 per ton for low-grade ore. They followed his advice and with minimal effort extracted a considerable amount of quality tin ore. It was around then Josiah heard of the death of Mr Humphrey Willyams, his very good friend and benefactor from Carnanton, Cornwall.

Having established himself as a reliable promoter of mining companies supported by reports that confirmed and validated his substantial experience, Josiah published in the business card section of the *Sydney Morning Herald* on 14 June 1872 an advertisement:

CAPTAIN J HOLMAN is prepared to inspect and report upon mineral properties. He is now in the neighbourhood of Glen Innes and Tenterfield. Communications addressed to the care of the Hon. SAUL SAMUEL 3, Spring-Street, will receive immediate attention.

NOTE: Saul Samuel was a senior NSW public servant, Colonial Secretary, the first point of contact for tenders and contracts. He was appointed Secretary for Lands 1867 and Sydney Chamber of Commerce, Vice-President of the Executive Council, NSW Treasurer with various governments from time to time and Postmaster General, 1872-1880. He was also a director of a number of mining companies including the Cadiangullong Consolidated Copper Mining Company, MLA for East Sydney, represented the Counties of Wellington (1854-56) and Orange (1859-60) in the new Legislative Council and was knighted in 1882.

Josiah had earlier recommended that Thomas Clark take over as manager of the Great Northern Tin Mines at Horseshoe Bend and William Blood would continue managing the Mount Mitchell Mine and store. In late June, Josiah boarded the Cobb & Co coach stopping off at Deepwater where he stayed the night at the Deepwater Post Office Inn. On a level area by Deepwater River, a number of buildings had been established and more were under construction. Streets had been surveyed and corner post marks could be seen where the intersection with these would meet. In the morning, Josiah bought provisions at the post office store, hired a cart and headed 15 miles to Bolivia, then west to Glen Creek just below Tent Hill. Here he examined a 60 acre lease known as McFayden's, owned by the Multum In Parvo Tin Mining Company. The area was weathered and rugged, holding quite a number of shallow swamps interconnected by small creeks which ran through the property. Studying where the streams intersected with the shallow gullies, he noted the extent of erosion which would indicate the possible extraction of minerals carried out over thousands of years. He was able to see that the mineral lode ran north-easterly and south-westerly and was easy to follow. The area had been surveyed earlier and an assay of a sample of unusually rich stream tin ore was conducted by Mr W. Twemlow and yielded almost 75 per cent pure tin. Having completed his inspection, he returned late in the evening to Bolivia and stayed at a very comfortable Inn. Crowded with miners, conversations focused on mining activities in and around the area.

In early September, the Multum in Parvo Tin Mining Company issued its prospectus to raise £12,000 via shares of £1 each. The company further stated: -

> The Company's ground (known as McFayden's lease), consists of 60 acres of well-selected and very rich Tin Land, on the celebrated Tenthill, Wellington Vale Run, adjoining the Banco Company's claims (Black and O'Keefe's leases), and is bounded on the south by Riley and Cohen's ground (M. Master's selection), the Great Britain Company's ground being a little to the westward of Lewis and party's selection).
>
> Besides being unusually rich in stream tin ore, yielding 75 per cent, of pure tin (see subjoined assay of W. Twemlow) the "Multum in Parvo" Company's ground is traversed by the four parallel "Banco" lodes, running north-easterly and south-westerly, these lodes being visible cropping out, and can be traced through the entire length of the, Company's ground, as per subjoined report from Captain, Holman, mining engineer,
>
> A creek runs through the centre of the property, and an abundance of water can be obtained from adjoining swamps. The wash dirt is easily got out, and the position of the ground affords great facility for immediate working. The quantity of ore on the Company's ground is inclined table, the four stanniferous lodes alone and independent of the large quantity of stream tin ore would give ample scope for constant work and large yields for very many years. Immediate and steady returns may be calculated upon soon after the

formation of the Company, as the tin can be procured without delay, and there is a good road the whole way to Grafton, via Glen Innes.

Soon as 150 of the 200 Promoters' Shares now offered to the public are sold, a meeting of the Shareholders will be called for the election of Directors, for consideration of the Deed of Association, registration of the Company, and such other business as may be considered necessary. The ground having been surveyed, the plan of the Government Surveyor, and also the samples of the ore, can be seen at the offices of the Company's Brokers, where applications for the Promoters' Shares will be received.

Josiah's report was included: -

Sydney, 2nd September, 1872.
"McFayden's Tinlands, Tent Hill, Glen Creek. Having examined (about eight weeks ago) – some Tin Lodes situated in lands adjoining the above. I hereby certify to having traced two of the main tin lodes from Messrs. S. Black and O'Keeffe's leads into McFayden's leases. These lodes can be traced throughout the whole length of the latter lands into those of M. Master and Co.'s, and I am aware of the existence of two other parallel lodes running into this ground, but had no time to trace them.

From the well-known prolific tin bearing qualities of these lodes, seen both at the outcrop and in tests by shafts in Blacks and McMaster's selections. With McFadyen's lands situated in the centre, and having the lode outcrop equally large and well defined, there is every indication to warrant the assumption that the lease here will when judiciously developed prove very remunerative and inexhaustible. These lodes have a bearing of S.S. W. and N.N.E., end underlay went into this selection.

The rock enclosing these lodes is granite, and the selection is situated on high tablelands well covered with valuable timber for mining purposes. As a whole I recommend the mine as a good and profitable investment. "

Josiah Holman, Mining Engineer.

While in Deepwater in early July 1872, Josiah received a letter from John Frazer, a director of the New South Wales Moonta Copper Mining Company to undertake a mine assessment prior to the company going public. Frazer, along with Saul Samuel, was also Director of Newstead Tin Mining Company near Inverell and had been introduced to him the previous year by Corbett Lawson. Josiah was familiar with the company as they had used his smelting works at Cadia for a number of years, selling the pure copper through Richard Morehead's SAIC.

From Deepwater, he travelled to Inverell and spent some time examining the Newstead mine works. It was reported in July 1872, the Newstead Tin Mining Company had shipped seven tons of ore to London via the 2044-ton *HMS Galatea*. Over in one gully, W Jenkin, the mine manager had washed out as much as 60 lbs of ore to the load, however, lack of water for sluicing continued to plague operations. Over 12 tons of streaming tin had been mined and what was required now was heavy machinery for crushing. A "competent practical engineer" was required to select the most appropriate site and map out plans for its construction. Josiah spent almost a day at the site as well as inspecting the two new shafts, now down to 60 and 24 ft. The *Armidale Express* on 13 July 1872, writing of Inverell posted:

> Captain Holman, I see, is again amongst us. I think, when he has visited our different mines now in active work, that he will not have so very high an opinion of the Oban mines in comparison as he held when he left us last time. When he has been on Cope's Creek, and has seen three or four men turn out half a ton of good clean-washed ore for their day's work, he will, I imagine, form a better opinion of us than he had from his first flying visit.

Having spent a number of days examining and appraising Saul Samuel's Darby Creek mine and those along Cope's Creek, the following morning Josiah caught Low's four-horse coach from Inverell to Armidale and then the Royal Mail coach stopping at Tamworth where he stayed for the night. The following morning, he headed east 12 miles to the small village of Nundle.

He was warned about the condition of the road, a grievance of the miners, so chose to travel by horse from the local livery. It was a steep descent before reaching the river flats, where he continued on a mile or so to Oakenville Creek to examine new mineral finds which had been uncovered. Gold had been discovered a decade earlier at Swamp Creek and as news travels fast, it attracted a wave of prospectors eager to seek their fortune. The once-small village had now grown to over 500 and substantial brick buildings were being constructed. After crossing a sturdy bridge over the Peel River, he came upon a fine hotel, a court house and police station and residence along with scattered cottages. But the main activity was a further seven miles downstream at Bowling Alley Point where diggings scattered the valley flats and races, some as long as 15 miles spread like fingers in various directions. The usual shamble of huts marked the boundary of a road of sorts along with a number of stores, a post office and four hotels. The large wheel of the crushing mill worked tirelessly, when water prevailed, next to obligatory stampers and a little further on a horse, more dependable could be seen on the slope working a whim. Numerous claims in the area had been made already as trial shafts had rewarded some lucky miners who had discovered rich reefs. Shares in the established Golden Streak claim had been sold at up to £150 each. Josiah spent some days examining the area and believed the claim offering the most likely potential was Foley Folly, known as Prisk's. It was rumoured, that after discussion with the owner, Josiah offered £10,000 for his claim, but this was refused.

From Nundle, he returned to Tamworth and caught a coach to Scone, then to Cassilis, Mudgee and Bathurst via Sofala. After staying at Bathurst overnight, he caught the early morning Cobb and Co. coach to Cowra and continued a further 12 miles to a mining site which covered 120 acres of which 40 were freehold and 80 leased from the Crown. It had previously operated as the Belmore Copper Company, but due to lack of capital had ceased operation. The high cost of cartage had drained all additional funds, thus not allowing for adequate testing of the mine. The directors now wished to raise capital by selling 100,000 shares at 10 shillings each to purchase the Belmore Copper Mining Company and further develop the mine following Josiah's recommendation. Three shafts had been sunk, the deepest 240 ft on the southern side of the 40 acre block some years earlier. A very rich lode of red oxide of copper, some 6 ft wide, yielded 84 1/2 tons of ore which had been smelted at Cadia and produced over 21 tons of pure copper. Prices were then £71 per ton but by 1872 had risen to £110 per ton. As the opening of the outcrop of the lode was on the side of a hill, Josiah recommended driving an adit at a horizontal level to meet the 40-ft shaft and then extend it further into the lode. He also recommended additional shafts be constructed where other outcrops of ore had been discovered in adjacent blocks. Having completed and posted his report to John Frazer, Josiah returned home to Cadia.

In the business card section of the *Sydney Morning Herald* on 30 July 1872 an advertisement appeared:

TO MINING COMPANIES. -The undersigned has proceeded to Orange, where he will remain for a short time, and offer his services to inspect and report upon MINERAL PROPERTIES in the Western district. Reference kindly permitted to Hon. Saul Samuel, 3, Spring-street, Sydney. JOSIAH HOLMAN, Orange.

On 31 July 1872, a prospectus notice was listed in the *Sydney Morning Herald*.

PROSPECTUS of the New South Wales Moonta (Late Belmore) Copper Mining Company (Limited), Cowra.

120 Acres of the finest copper-bearing land in this well-known district.

Capital £50,000 in 100,000 shares of 10s each.

50,00 shares and £3000 reserved to the proprietors for the purchase of this very valuable property, and 60,000 shares offered to subscribers PAID-UP TO 5s EACH on the following liberal terms: -1s per share on application, 1s 6d per share on allotment, and the balance, 2s 6d, in calls of 3d per month if required.

Dividends payable on all shares alike.

It will thus be seen that a payment of 2s 6d. and prospective calls of another 2s 6d per share (or 5s in all) will entitle each share to rank as FULLY PAID-UP.

Directors: John Frazer, Esq., JP., W. J. Watson, Esq. (Watson Brothers, Young.) Alexander Brown, Esq., City Iron, Works, William Bollard, Esq.

Broker to the company Josiah Mullens, Esq.

The Proprietors pay all expenses connected with the floating of this mine. This Company is formed for the purpose of purchasing and working the New South Wales Moonta Copper Mine, situate within 12 miles of Cowra, and known for many years past. as the Belmore Copper Mining Property, comprising 120 acres of copper bearing land, of which 40 acres are freehold and 80 acres leased from the Crown in blocks of 20 acres each.

The New South Wales Moonta mine was discovered in 1858, and was worked for a considerable period up to its present success. One smelting of 84 tons ore produced 21½ tons pure copper at the Cadia works, and sold in England through the agency of Messrs. Morehead and Young, as per account sales in the hands of the agent of this Company. This mine is well and favourably known in the district, and the surrounding land for three miles in extent on either side has been selected. A great number of shares have already been applied for by local residents.

The operations at this Mine were discontinued, owing to want of capital and the low price of copper then ruling in England. The proprietors, without any desire to over-estimate its value, merely state they place this investment before the public with every confidence that it will be, within a short period, one of the best dividend paying copper mines in the colony.

Several parcels of ore have recently been tested at the Mint and elsewhere, and yielded from 40 to 58 per cent of copper. Abundance of timber and excellent fireclay are on the ground; carriage to Sydney is estimated at £4 per ton. Attention is directed to Captain Josiah Holman's report appended hereto. J. Mullens, Broker to the Company.

Josiah's report, in part, reads as follows:

I feel no hesitation in stating that rarely has a richer and most prolific bunch of copper ores been extracted than your mine has produced, and as the chief works requisite for its early development will consist of adits, a very moderate outlay will suffice. The mine is easy of access at all points, and the immediate district affords excellent and abundant supplies of timber for smelting and mining purposes. Specimens of these ores, now on view, will favourably compare with those from the richest mines in these colonies.
I am, Sir, yours obediently, Josiah Holman, Mining Engineer.

On 11 July, Forster, Kelly, and Co. reported in the *Empire* newspaper, that the Peabody Copper Mining Company, near Molong, 30 miles north of Orange, had been placed on the market, at 10 o'clock yesterday morning, and the office was rushed with applications. At 3.15 p.m. the application list closed with several thousand applications, well above the requisite number, including a large application from Orange. The shares immediately advanced in premium. A telegram had been received, saying that samples taken from the main shaft at a depth of 36 feet, showing ore of extraordinary richness were being forwarded by coach. Five days later the paper reported that shares had traded as high as 3s 6d, 3s 9d, and 4s and yet, a prospectus was still waiting to be released.

Such were the unprecedented speculative times in the colony. There were days when more than two full pages of newspapers were filled with notices of impending company floats or details announcing a new extraordinary mineral find. Newspapers described brokers as being "besieged" when offers were announced and it was reported that over £25,000 had been withdrawn over three months from the Government Savings Bank. It was as if every square mile of the country had been pegged out to lease. It would be almost a century before such an extraordinary mining boom would again be experienced in Australia. On the flip side, the immaturity and inexperience of the young colony was exposed. The Mount Mitchell Tin Mining Company, managed by Captain Holman, the Caledonian Tin Mining Company, Captain Edwards manager, the Oban, Captain Williams, manager and the Great Southern, all limited liability companies with a nominal capital of from £30,000 to £60,000 had yet to complete preliminary formation.

A correspondent writing for the *Sydney Mail and New South Wales Advertiser*, on the 13 April 1872 illustrates it best:

> If rumour speak truth, and the prophets upon Change do not prophecy falsely, New South Wales is upon the eve of her Golden Age. The spirit of speculation has seized upon her colonists. Impulsive persons, eager to make colossal fortunes, and too impatient to plod along the safer and more beaten paths of trade, are daily and hourly plunging into a wilderness of quartz reefs on the spec of finding short cuts to wealth.
>
> Men with money to lose are staking their bird in hand upon the prospect of catching unknown numbers of birds in the bush. Men with no money to lose are picking up the stakes. Merchants, squatters, farmers, shopmen, clerks have rushed with headlong impetuosity into the share market. Shares and scrip, and calls and dividends, and claims and leases, are now the staple of conversation.
>
> Adventurers long since overwhelmed beneath the tide of adversity have come suddenly to the surface, and are now to be seen actively engaged in the apparently remunerative occupation of floating companies. Seedy men, who have some mysterious and incomprehensible means of inspiring the public with confidence in their honesty, lounge about the haunts of trade, exhibiting fragments of gold-besprinkled quartz, descanting upon the richness of this claim, the prospects of that, and the vast fortune which will shortly be at the disposal of the lucky fellow who acts upon their advice.
>
> An army of sharebrokers has risen up as suddenly as the mountaineers of Rhoderick Dhu[*], and now fully garrisons Greville's Rooms. The columns of the *Sydney Morning Herald* are full of prospectuses — encouraging prospectuses, which tell of wondrous wealth, and how easily it may be won — craftily worded prospectuses, each one professing to give the only true directions as to how an El Dorado may be reached. All is bustle and

---

[*] Rhoderick Dhu — Black Roderick — a villain in Sir Water Scott's poem *The Lady of the Lake*

excitement. The traditionary slowness which has been made a reproach to Sydney has vanished, and everybody one meets is an eager speculator.

In early August, Josiah was approached by the Peabody Copper Mine north of Orange, near Molong, to carry out an extensive appraisal of their mining lease. Two mining captains had already visited and examined the new workings. Captain W. Reynolds of Icely copper mine nearby examined a mineshaft sunk to 4 ½ fathoms and reported:

> The veins increase in richness and solidity as they descend … All the ore I examined was of a much higher percentage than any I have seen before.

Carrangera copper mine Captain William White stated:

> The rock is principally amorphous limestone traversed from north to south by a dyke or load of calcite, with the width from 3 to 6 feet. In this dyke or load the ore or native copper is found. It is traced at the surface by shafts and prospecting pits a distance of 160 feet and in all of them I got out exceedingly rich specimens of ore and native copper.

As the site was 50 miles from Josiah's home in Cadia, he was of course very familiar with the mineralogical landscape of the area. His report dated 20 August 1872 read as follows:-

> To the Directors of the Peabody Copper Mines, Sydney.
> Gentlemen, Having recently seen exhibited in Sydney some remarkably rich specimens of red oxide and native copper, the product of the Peabody Copper Mine, situated near Molong, and having to make a tour in that district, I thought the Peabody offered sufficient inducement for my diverging a little of my business track to view, for my edification, this unusual phenomena in copper mining in New South Wales.
>
> Arrived at the mine, I at once recognised heaps of amygdaloidal trap rocks excavated about the shafts, precisely identical with the formation in which

the Americans during the last thirty years have raised such immense masses of native copper, and in such unprecedentedly wonderful quantities as to astonish the geologists and mineralogists of the day.

In 1854 I visited several of the most celebrated native copper mines on the southern shores of Lake Superior, United States of America, and having since travelled very extensively among mines, it is singular that, till my visit to the Peabody, I had not come in contact with amygdaloidal trap similar in all its characteristics and associations to that of the Lake Superior districts. My surprise at the production of native copper at the Peabody was smoothed down by the fact that there exists in the Molong district, a dyke of rocks that coincides with the Lake Superior amygdaloidal traps, and which will, doubtless someday, when developed, give masses of native copper similar to those hitherto produced is such huge blocks only in the North Americas continent. In the Peabody mine, the native copper and red oxides of that metal have been traced along the surface, in a north and south course, fully 100 yards in length, and at the deepest point attained-some 36 feet deep I saw fine lumps of metallic copper dug out.

In conclusion, I would remark that the rich patches of copper already developed fully warrant greatly extended trials being made to develop this extraordinary mine.

I am, Gentlemen, yours truly, JOSIAH HOLMAN, Mining Engineer.

Sydney

August 20, 1872

In early December, Josiah travelled to the Wellington District to inspect the mineral selections held by Joseph Weame. Weame was a member of the Legislative Assembly, Member for Central Cumberland, and a director of The Pride of the West, No1, Quartz Company, Cheshire Creek, located some 15 miles north of Bathurst. At the end of December, the mining manager, reported;

I have not seen a more encouraging prospect during my eighteen years' experience in reefing than this 20-acre Lease, for the work that has been done. I started our men to open out on our northern boundary, and struck the reef at a depth of four feet. Gold to be seen in nearly every stone in fair quantity. I intend putting in a tunnel about eighty yards down the hill, and driving on the course of the reef up to the boundary. This will prove the reef as we go on, and will give us 200 feet of backs to work to the surface. The men are now busy in shaping the drive. I expect to cut the reef in a few feet from the mouth of the drive. In cutting a trench to carry the water from the mouth of the drive, we came on some quartz near the back of the reef coated with gold. Captain Holman will bring you a sample.

Meanwhile, all seemed to be running well at Mount Mitchell under the supervision of William Blood with over three tons of tin ore being shipped to Murrurundi Railway Station in January 1873. Emily and William had their second child, William Fredrick (Fred) born on 29 March 1873 in Oban. However, the dispute over shares and distribution that had been simmering for over 12 months was finally put to a meeting of shareholders of the Mount Mitchell Tin Mining Company on 7 May 1873. Unfortunately, the proposed resolution was not agreed by all and legal action commenced on 28 May. A meeting was called to consider winding up the company on 27 August but adjourned to 26 November. On Saturday the 3rd January 1874 an auction was held to dispose of:

> All the Valuable PLANT of the Company, consisting of mining tools, of all descriptions for mining, carpenter's, and blacksmith's work of the Mount Mitchell Mining Company.

William Blood, Emily and family stayed on at Paddy's Gully in the house Josiah had built, managing the now almost defunct mine until 1875 when their third child, Henry Earnst was born in Armidale. They returned to Cadia around late 1877.

Prices of tin, holding steady for a number of years at around £100, continued to suffer with the opening of new mines in Malaya, Burma, Bolivia and Australia and had fallen to less than £40 per ton.

On 5 May 1873, Newstead Tin Mining Company called a special general meeting of shareholders to "consider the present position of the Company's affairs".

On 16 April 1874, the company postponed to 23 April a special general meeting to ... consider the advisability of winding up the Company.

On 25 September 1876, a special general meeting was held and a resolution passed:

That the Company be dissolved, and the assets realised.

On 16 January 1877, Newstead Tin Mining Company shareholders were advised that:

A return of capital of one half-penny per share upon all shares.

In August 1873, a special general meeting of the Peabody Copper Mining Co. (Limited) was held and a resolution passed to wind up the company. Protracted legal claims dragged on until January 1877 when the Court ordered a refund of 2d. per share be paid.

In November 1873, a notice was given of a meeting to be held on 10 December to consider dissolving and winding up the Great Northern Tin Mining Company. On Tuesday 21 April 1874, the assets of the wound-up Great Northern Tin Mining Company were distributed at their offices, 5 Spring Street Sydney, at the rate of 7 ½ d per share, on contributing shares only.

The mining history of New England has been repeated all over the world. The glitter, like a light that attracts moths, lured an eclectic assortment of fortune hunters, all with one purpose, securing untold wealth. Oban is no more, just one enlightening example revealing mining fickleness, great dreams, excitement, then quickly deserted, abandoned, the wealth mere memories deferred to history books like this.

Peak Downs Copper Mine

1873-76

While Josiah was making his way from New Zealand to Cadia in 1862, John Manton, manager of Colonel O'Connor's mine just south of Gladstone, was negotiating with two Sydney entrepreneurs, Thomas Sutcliffe Mort and Thomas Ware Smart, to establish a company to raise money to develop a new mineral find. Around the Peak Range in central Queensland, miners seeking their fortune came from far and wide in search of gold and were scattered about the creeks and valleys in the hope of finding riches. John Manton had travelled west from Rockhampton and while exploring, met one such miner, Jack Mollard seeking his fortune. In his search for gold, Mollard had come upon a hillside containing another mineral he thought may be copper. Manton's extensive mining experience told him it was most likely copper and offered to buy Mollard's mineral rights. Surveying the *"mountain of copper"* he immediately made claim for three allotments totalling 240 acres just south west of Clermont. Samuel Stutchbury, a mineralogical surveyor appointed by the NSW Government had visited the area almost a decade earlier and stated:

> During the whole of my travels, I never came on a spot so full of minerals. Turn my head where I will, I can find copper, lead, silver, and gold; and if I regret any thing in my life, it is that I am not able to carry out my survey, the Government breaking up the party in Gladstone.

Manton's mining experience also told him that there would be massive challenges ahead in mining this property successfully. The nearest port was

Rockhampton some 250 miles away; the area was rugged, often semi-barren and perilous for travel. Roads, where they existed, were quite treacherous, as he would soon find out when the severe wet season at the end of summer would make them almost impossible to navigate. Droughts would wipe out any feed that might be available for bullocks, thus requiring transporting of feed, further driving up cartage costs. This countryside was far from hospitable and was once described as being suitable only for mining. It was a far cry from its rich cousin further south, the Darling Downs with its pleasant rolling plains. Attracting a capable, experienced mining workforce was likely to pose yet another formidable challenge.

To verify his initial assessment, in August 1862, Manton secured the services of two experienced geologists Arthur Ledger and Pedro Nissen, who with James Scott the Government Surveyor, travelled to survey the area just south-west of Clermont. Five specimens were taken from different areas and sent to Professor Charles Watt of Sydney University who reported the specimen's contained yields of between 17.6 and 33 per cent copper. Mort and Smart hearing of this discovery and its very positive results, claimed two additional blocks bordering Manton's claim. With no funds to develop his claim, Manton agreed to join with them. It was agreed they would float a company to raise money to develop the mining lease and, in the interim, Mort and Smart, using their own capital, would engage miners to begin the mining operation. A notice appeared in the newspapers on 14 February 1863 announcing the formation of the Peak Downs Copper Mining Company (PDCMC) offering 100,000 shares of £1 each.

Captain W. Burgoyne was appointed mine manager and set about drawing up plans to follow the exposed lodes and within six months there were over 50 miners at work on the site extracting ore to be taken to the coast. With a number of mines being closed down, miners came from far and wide, California, Burra Burra and Cornwall. But not all stayed, many left Peak Downs attracted by the promise of riches in new gold discoveries at nearby Clermont, their places being filled by less reliable men of whom Mort wrote: "Great have been our disasters owing to the drunkenness of our employees."

Mining was initially mainly superficial with only two underground shafts being developed due to the lack of suitable hard-rock mining equipment. Recognising one of the major impairments to success was going to be the cartage of the ore some 250 miles to port, on 28 May 1863 they advertised for tenders prepared to carry supplies and equipment from the port to the site and return with ore to the port.

> WANTED — TEAMS for Ten Tons Loading to the Peak Downs Copper Mine Apply to Shaw, Copper, & Co.

Advertisement 13 May 1863

A round trip took more than two months — if the road was in fair condition. It was hoped a carrier could be retained at a cheaper rate than the current £30 per ton being charged. In South Australia, cartage was almost a third of that. Almost impassable by road in droughts and inaccessible in floods, until the transportation issue was resolved, financial prosperity could never be achieved. Undaunted by such obstacles, areas were cleared, plans drawn up and tenders called for supplying 200,000 bricks for the planned five furnaces in anticipation of the furnace equipment due to arrive from London around September. They estimated producing at least 150 tons of copper each month.

Incorporated on 13 November 1863, the company issued only one-third of the 100,000 £1 shares, the balance being reserved by the vendors in the proportion of half to Manton and one-quarter each to Mort and Smart for providing the initial working capital. With the £30,000 capital raised, pace at the mine site increased. When news of the continuation of the rich lode reached Sydney, speculators rushed in and the £1 shares issued months earlier quickly rose to £4. Their rise was as spectacular as the rivers and creeks after a particularly severe wet season drenched the north of Australia in early 1864. Predictably, this created havoc in the mines and brought carting to a standstill. Rumours of mismanagement and over-estimation of ore reserves seemed confirmed when Manton, the original owner, was dismissed from the position of general manager.

A more experienced manager was required and Captain Josiah Dennis from the Burra Burra copper fields was appointed mining captain in April 1864. He would be ably assisted by Mr Honey, accountant and superintendent, to implemented more stringent, regulatory reporting procedures and directives to instil a higher level of discipline and workmanship. Having explored the area further, Dennis began construction of a new shaft, Thomas shaft 30 yards from the main lode and at the 17 fathoms level struck a very rich lode over 4 ft wide. Another new shaft, Williams, excavated further into the rich lode also achieved success, in all raising over 2000 tons of ore at a yield of almost 25 per cent selling for around £90 per ton. Unfortunately, this hardly covered the operational costs and added to that was the construction costs of two furnaces. No dividend was declared.

> TO BRICKLAYERS and BRICKMAKERS.
> —Constant Employment, at highest rates of wages, is offered to steady men at the works of the Peak Downs Copper Mining Company. Agents in Rockhampton, Messrs. SHAW, CAPPER and Co., by whom all information will be given.
> 4076   GEORGE SPAIGHT, Secretary.

The Brisbane Courier Wednesday 5 April 1865

In and around the property, the clay was proving to be excellent for brickmaking and development of the refinery began once enough bricks had been fired. While the mining operation was developing satisfactorily, the smelting operation was forced to cease. Smelters imported from England faced challenges not experienced before, their lack of skill laid bare by the results. Isolation added to the challenges of experimentation, wood for fuel, clay for hearth bricks, sand for smelter bottoms, flux for chemical reaction overwhelmed the less experienced. Too much blister copper was sent to Swansea where brokers were demanding exorbitant fees to sell the inferior copper. Shipping costs and brokerage substantially reduced the Peak Downs company's returns and it was decided to suspend that method of operation until more experienced smelters from South Australia arrived. Continual difficulties transporting the ore to port to be shipped for smelting not only increased costs, but also impacted the revenue stream, forcing the company to apply for an overdraft from the Bank of New South Wales. Then, in mid-1865, George Speight, company secretary, disappeared. Concerns were raised with news he had boarded the *Guipuzcoano* for Callao Peru on 19 July and absconded with a reported £400.

When the smelters arrived from Burra Burra on 30 September 1865, they immediately began reassembling and reconstructing the furnaces utilising Welsh reverberatory technology. This put additional delay on resumption of smelting and, added to that, it was soon discovered local inferior sand for the furnace bases was not well suited to handle the immense heat required for smelting. After searching as far wide as Sydney, the smelters finally decided to transport, at considerable cost, sand from Adelaide that was already proven. Having completed correcting the mistakes in the initial smelter construction and operation, the men began developing three additional furnaces. They also completed a manager's residence, office and huts for the miners, store and stockyards to keep cattle. Despite that, many of the miners and families still lived in tents and others preferred to live in the nearby townships of Copperfield and Clermont. By the end of the year, the creeks, valleys and low hills around Clermont were supporting almost 3000 seeking their fortune in one way or another. The hot humid summer months and tropical downpours continued to take their toll on the miners, although the Chinese, of which there were many, seemed to handle the wretched conditions much better. Chills and fever plagued those unable to secure quinine and other medication.

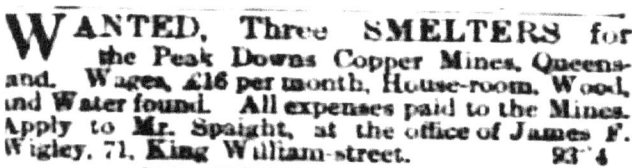

South Australia Register 3 April 1865

The appointment of general manager William Woodhouse, mining captain Josiah Dennis and smelting captain David were having a positive impact. Construction of reservoirs for the conservation of a water supply were undertaken and an engine for hauling as well as pumping water to dress the ore had been ordered. Additional miners, mechanics, carpenters and a blacksmith had been added to the payroll and construction of cottages and huts was ongoing. With the addition of 50 miners from South Australia, Captain Dennis set about implementing plans to sink a shaft 30 fathoms and then construct a crosscut at that depth for 20 fathoms through the lode. Additional competent smelters and furnace masons had been engaged to assist the building of 10 furnaces capable of producing the 200 tons of copper each month that Captain Dennis was hoping to achieve. This was far less than the output of the celebrated Burra Burra mine producing over 4000 tons in six months.

With the smelting problems mostly resolved, a fourth smelting furnace now in operation and a fifth almost completed, bullock wagons left more frequently, slowly removing over 1200 tons of ore on hand — in total estimated to be worth over £17,000. Shifting the huge backload of copper certainly helped the company's revenue stream, but later in the year prices gradually fell to from £100 to £80 per ton. On its way to Sydney were 14 tons of copper, rough copper securing £80 a ton in London. Continuing concerns over the 250 mile cartage to Rockhampton seemed set to be resolved with plans for improvement of 125 miles of the road to the port of Waverly in Broadsound and construction of an upgraded port facility. The round trip from the mine site to the port, Waverly, and return took almost a month if conditions were favourable. Not surprisingly, with the huge increase in the payroll cost and capital expenditure, the company ran out of money and again no dividend was declared.

On 8 March 1867, John Penrose Christoe travelled from Sydney to Brisbane on the 536-ton *Yarra Yarra*, then boarded the 425-ton *Lady Young* for Rockhampton, where he caught the coach to Peak Downs. Arriving in late March, he took over management of the smelting department as head assayer and through the next few months was joined by a huge influx of miners, smelters and their families from South Australia. All were leaving over concerns that the rich Burra Burra mine was almost exhausted. Around the same time, the machinery for pumping and hoisting arrived. With new equipment to help with mining and processing, ore excavation substantially increased. Almost 2500 tons of ore was extracted, averaging 20 percent copper during the final six months of 1867. All this was helped by the seven furnaces now in operation and would be further assisted once the three under construction were completed. In total, almost £100,000 had been spent on the development of the mine, its management and production of copper. The upgrade of the road to Broadsound had been completed, the journey now taking 21 days instead of a month, thus reducing cartage costs to £8 per ton. Finally, the long-suffering shareholders were rewarded when a dividend of 8 per cent was declared, even though the overdraft, attracting a high rate of interest, remained.

Captain Dennis soon had the new arrivals working productively, adding new shafts to the labyrinth of adits, drives and stopes under construction; over one mile was planned for further excavation. When one of the new shafts (stope) struck a rich ore lode averaging 55 percent copper, excitement rippled throughout the mine site. This euphoria was short lived as a severe drought took hold of northern parts of Queensland halting cartage for a number of months. Completion of the 14 furnaces and two refineries was delayed and lack of fodder for the horses meant ore could not be raised from the deep shafts. Even so, more than 7800 tons of ore was extracted and the furnaces produced more than 1000 tons of copper ingots and cakes with an average just above 20 percent copper. But prices had continued their steady decline, agents in London being only able to obtain £75 per ton. However, all this did not prevent a 10 percent dividend being declared.

In early 1869, Leyshon Jones from South Australia was appointed smelting manager. Christoe remained head assayer although Jones's stay was short lived. He left after less than a year, replaced by Maurice Thomas who had previously worked for the Burwood Smelting Works, Newcastle. In March

1869, the general carrier and forwarding agent company, Woods, Shortland, Fox, and Co, entered into a four-year agreement to transport ore, ingots to port and return with mining equipment, goods etc, at the rate of £10 per ton. They would require 250 horses, 50 sets of harnesses and wagons, each to carry 2½ tons of ore. All 14 furnaces and two refining furnaces were now operating 24 hours a day, seven days a week and the mining equipment that had arrived earlier was now operational, the huge steam engine driving the wheel and connecting wire ropes and pulleys lifting the ore from the shafts. The new management team were certainly kept busy managing the 400-strong workforce engaged in mining, carting, wood-cutting, refining, washing, carrying and all manner of mining activities. In each of the six shafts, levels were being driven horizontally into the lode every 10 fathoms down to 40 fathoms and in other areas, a connecting series of tram roads with their small carts had been constructed both below and above the mine intersecting with the smelting operation. With the additional miner tents and recently built cottages and huts, the vacant area around the mine site was slowly being transformed into a small village. A dividend of 15 percent was declared for the year. Although it had taken a number of years before dividends had been paid to shareholders, those who paid £1 when the company was floated had by now almost doubled their investment with dividend payments. Christoe did not stay long however, and left when offered a position by Morehead as smelting manager at the Newcastle Smelting Works.

The monsoonal wet season in north Queensland can be drenching and such was the case again in early 1870. Over 100 drays returning with equipment and goods or heading to the coast laden with copper ingots and cakes were either bogged axle deep in the thick black soil or held up waiting for the flooding rivers to recede. For the teamsters, the few weeks to reach the coast became a few months, in some instances almost six months, a long time on the road. The smelting operation was severely affected. Building and repairs were hampered by the continual rain and fuel was running out as carters could not return laden with timber. Added to that, goods ordered for the mine could not reach their destination and when the mine ran out of candles, all underground mining excavation ceased. Creative planning of the mine drainage layout, however, saved the mine from potential disaster. They were lucky enough to be able to excavate almost 6400 tons of ore during the year, the yield being maintained at just over 20 percent. When the rain finally stopped, it took almost six months to clear the backlog of copper ingots and cakes from the Peak Downs mining site.

Captain Dennis continued his search for rich copper veins and shafts were sunk deeper to almost 50 fathoms where they encountered harder rock and unfortunately poorer quality ore. Rather than continue, it was decided to sink new shafts, which would also assist the drainage of the mine. During 1871, with better conditions, over 12,000 tons of ore were extracted and almost 2500 tons of pure copper left the north Queensland ports. But with the smelters running at full capacity, timber near the site was slowly depleting forcing carriers to travel further for timber, thus raising prices for the 300 tons of fuel being used on a daily basis. Encouraged by this activity and confident copper prices had bottomed, depressed for almost three years now slowly rising, a dividend, unheard of before, of 80 percent was distributed to shareholders.

As new mineral areas were discovered and large-scale operations were beginning to be developed, population in the area increased, many taking up land holdings for cattle or cultivation. Politicians of many persuasions along with captains of commerce made continual and vocal representations to the government to build a railway line to the Clermont mining area. The anticipated huge reduction in cartage costs would benefit individuals seeking their fortune in the already well-established mining operation at Peak Downs as well as those involved in other pastoral industries such as wool and sugar.

The challenge of securing a large, competent workforce continued to disrupt the mining operation, even though new personnel were being hired. There were now almost 600 workers, more than 100 of them underground miners. Almost monthly, advertisements appeared for smelters, furnace masons, bricklayers and miners, 100 if they could find them, willing to take the trip to Rockhampton and then a few weeks over rugged terrain to Clermont.

Through their London agents, 96 Cornish underground miners were hired and with their families left Plymouth and then Gravesend on 10 April on the 1319-ton barque *Indus* under Captain Ellis for Brisbane via the Cape of Good Hope, arriving on 3 July 1872. That evening they boarded the steamer *Black Swan* for Broadsound, arriving at the port of St Lawrence on July 9 before heading west to Clermont, their new adventure.

> **To Miners.**
>
> WANTED MINERS at the Peak Downs Copper Mines. Apply at the Works, or to
>
> GEO. B. SHAW,
>
> Agent P.D.C.M. Co., Rockhampton.

Advertisement 9 September 1871

A new more powerful engine, with all the necessary pumping and winding gear had been ordered and arrived at the mine site in early 1872 during the sinking of the deep shafts. Copper prices continued to rally and by mid-1872 over £100 was being paid per ton for quality copper. Now, with 21 furnaces working day and night and the men on eight hours shifts, a record 1450 tons of copper was produced during the six months period. The £1 shares that had languished at less than half that price for many years jumped to more than £4 and continued rising. But when copper prices reached their peak at £110, they soon began to slowly fall and by the time the ingots and cakes reached the markets in London, were trading back down at £80 per ton. Such were the vagaries of copper prices worldwide. The huge dividend paid six months earlier had been determined based on selling copper at £100. This shortfall severely impacted cashflow and left the directors red-faced; no dividend was declared. Added to that, the company was now saddled with a sizeable debt, nearly £42,000, the onward effect crippling future development. Any funds available needed to be used to pay the earlier declared dividend and consequently development of additional shafts recommended by Captain Dennis were abandoned. Furthermore, as the company was not incorporated under the Limited Liability Act, many shareholders, fearing that its losses would implicate them, endeavoured to sell at any price. Shares, which had been looked upon as good value for £8 to £9 each, were now changing hands for as little as 5s to 2s 6d each.

The challenges of climate continued to impact the mining and cartage operations when once again, summer tropical monsoons caused severe flooding in Central Queensland. Impassable roads choked by weeks of unrelenting rain prevented the collection of timber and eventually, one by one the smelters were closed. But worse, goods that were needed by the large workforce and their families meant the mining camp was quickly running out of food, leaving the remote mining site on the brink of starvation. At one point, they had completely run out of flour, a staple necessity. It was reported that one enterprising woman was able to reach the mining site carrying many bags of flour which she sold for £6 per bag. For a miner, that was equivalent to two weeks' wages. Not used to such conditions, for a number of the new Cornish miners and their families, the hot wet humid conditions proved fatal. With all future operations brought to a standstill, a feeling of depression set in, deepening, as time went on verging into almost panic. An unfavourable report from Captain Dennis could have not come at a worse time. The quality of the ore from the new ground already opened up had begun to deteriorate and there was great difficulty in raising anything like the previous amount of ore. At a significant cost, Captain Dennis sunk a shaft to a considerable depth without discovering any mineral of importance. A further distraction for Captain Dennis was a lawsuit with one of the timber carters.

None of this of course would impact the hundreds of miners engaged by agents in London already on the open ocean on their way to Rockhampton. The *Polmaise*, 753 ton, left Plymouth on 16 June 1872 and arrived in Hervey Bay on 12 September with 33 miners and their families, around 60, who were then transported via the *Tinonee* to Broadsound and Peak Downs. Another of these, the 965-ton *Countess Russell* — which had left London on the 4 March 1873 with 366 on board — arrived in Rockhampton on 30 June. Unfortunately, the vessel was required to be quarantined as a typhoid-like fever raged throughout the ship. When it arrived, 11 adults and five children had succumbed to the fever and more than 40 were ill. A number of those who were struck down by the fever were miners. More would die later. Dr Salmond, the health officer, visited the ship and instructed passengers to be quarantined onshore. In all, 36 people perished on that voyage. From the Peak Downs company directors' report of 11 August 1873:

"There are 562 men and boys employed on the mine, and 114 of the miners ordered from England have lately arrived at Rockhampton, in the Countess Russell."

These miners hired from England for Peak Downs left Rockhampton on the 25 July on the *Tinonee* for the port of St Lawrence 100 miles to the north of Rockhampton.

> COPPER MINERS. — Wanted, experienced Copper MINERS for the Peak Downs Copper mine. Apply at the Company's Office, 32, Bridge-street, Sydney.

Sydney Morning Herald 22 February 1873

Around June 1873, the company contacted Captain Josiah Holman and asked him to carry out a full assessment of the underground mining operation. As one of the most experienced mining captains in Australia, it was hoped he could provide some solution to ongoing problems that had plagued the mine for the past nine or so months. In the directors' report of 11 August, much was made of "... the over estimated value of copper and ores at grass" and "...showing an over estimate of the company's assets on 30th June, 1872, of £41,777.15s" — not an insignificant amount. Reference was also made to the working of the mine being "... far from satisfactory" with the hope that Captain Holman "... in making a plan of the underground workings, so that the directors may be enabled the better to judge whether or not the mine is being worked to the best advantage." Even so, 1774 tons of copper were produced during the first six months and at £80 per ton, earned the company over £140,000.

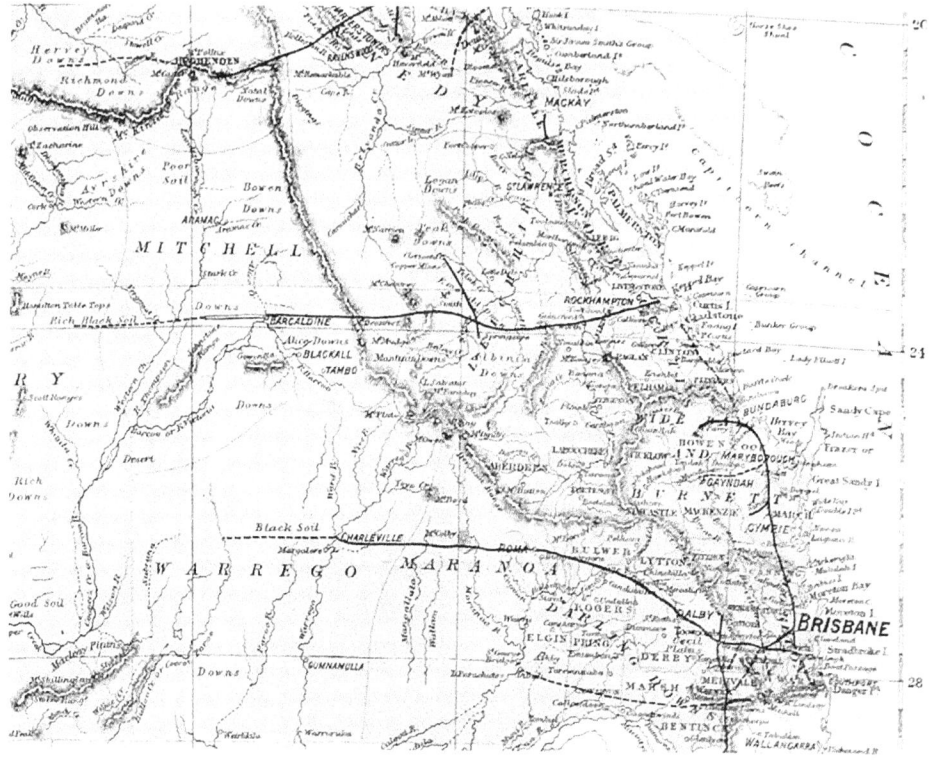

Queensland map

Having negotiated terms, Josiah left Cadia by early morning coach and caught the train from Kelso to Central Station. On 2 July, he boarded the 259 ton steamship *Balclutha* at Circular Quay bound for Rockhampton, arriving at the mine site in early August. Meeting with Captain Dennis, his initial effort was directed towards drawing up and mapping out the current and proposed underground works, many of the original drawings proving to be obsolete. Using these drawings, he was able to provide a comprehensive report on the overall operation. One immediate concern raised was the amount of ore being mined would not be enough to keep all furnaces operational. He was also highly critical of the reasoning behind the decision to sink one of the shafts beyond the lode. He directed the men to mine a cross-section and at the 18 ft level was able to locate a rich yielding lode. Of a number of the other shafts, he noted most had petered out; the initial high-quality mineral having been sustained for only a few fathoms before the payable lodes diminished into nothing. Inspecting the ore at grass, he determined most was of a poor quality, some only 6 percent copper, a considerable difference to that reported. Having taken almost three weeks, he completed his report on August 11 and then spent a few days catching up with some of the miners who had worked at Cadia as well as those recently from Kapunda where Dr Mathew Blood was practising. On 12 September, having returned to Rockhampton by coach, he boarded the Brisbane-bound 287-ton steamer *Queensland*, an iron, single-screw passenger/cargo vessel. He spent some time in Brisbane, which he reached on 27 September, before boarding the 504 ton *City of Brisbane* for Sydney. Arriving at Sydney Harbour on 3 October, he had time to redraw his sketchy plan of proposed works. Meeting with the directors in Sydney he discussed his report and after some negotiation, was offered the position of mine manager. This would be by far his largest and most challenging management role and a great test of his skills and experience. Peak Downs was then the second-largest mining operation in Australia. The famed Burra Burra mine was undoubtedly the largest venture undertaken in Australia, but by 1873 was being plagued with ore processing difficulties, its output having been reduced significantly. The next day, he boarded the train for Kelso and then a coach to Cadia via Bathurst and Orange.

There were possibly many reasons Captain Dennis resigned on the 11 August. Why did the board send an experienced mine captain to review operations? Were they not satisfied with his performance? Or did Josiah expose failings that would surely end his tenure? He stayed on at Peak Downs with wife, two daughters and son and left at the end of December when offered a position as manager of the nearby Mount Orange Copper Mining Company (Limited).

It had been agreed with Elizabeth before Josiah's departure in June that if offered a suitable permanent position in Queensland, she would travel with him. Not the best time to be travelling to Central Queensland with the impending wet season, but Elizabeth was eager to travel and given that Josiah had been away, on and off for the past five years she was keen to join him. As for the children, Emily Louisa was now married living in Oban, Joe (Josiah

Jnr), 20 years old and Charles William, 16 years old had finished school and were working on the property and assisting their father with mine operations. They were old enough to look after themselves and Annie was still at school boarding at Petersham.

Returning, Josiah and Elizabeth packed what they would need and set out instructions for what needed to be done and measures that had to be undertaken for the mining operation. They did not know how long they might be away. After two or three weeks, they caught the morning mail coach and headed to Sydney, catching up with friends and business associates before boarding the *Balclutha* for Maryborough and Rockhampton on 6 November. The trip was quite uneventful, the summer storms feared by many ship captains having not yet arrived. From the ocean, they made their way up the Fitzroy River, huge fig trees along the riverbanks giving a welcome reprieve from the vast ocean horizon of the days' before. Arriving at Rockhampton on 12 November, they unloaded their belongings and caught a coach to the Dunmore Hotel in William Street. They were able to spend a few days enjoying the bustling town before catching a morning coach to the end of William Street where they boarded a train to Westwood some 30 miles to the southwest. Awaiting to meet them at Westwood Station, was the 1.30 Cobb & Co. coach which took them through an undulating, hilly mountain range along a most uncomfortable corrugated road to Gainsford. Staying at the Gainsford Hotel, they were pleased the substandard accommodation was only for one night. At 6 o'clock on Sunday morning — the coach only ran from Gainsford Wednesdays and Sundays — they boarded the Cobb & Co. coach for Clermont. Once on the Downs, the landscape was not as heavily timbered, the farmers having cleared pockets for growing a variety of crops, but mainly sugar cane. It had not rained for a while and the horses and coach created clouds of dust which, if you weren't careful, could be very irritable. After resting a number of times to change horses, they arrived at Capella and the nearby mining site. The major township, Clermont, was a further 40 miles on, with stores offering a variety of items (in dry spells at least), a number of public houses —some run by Chinese gentlemen — and a bank, newspaper office and police station.

Around October 1873, a European geologist of repute, Rev. J. E. Tennyson Woods visited a number of mining sites in Central Queensland. He though Peak Downs copper mine "remarkable because distinguished by the immense quantity of rich ore which had been obtained from the mine, and also for the peculiar character of the ore, which differed from the copper ores found in Victoria and South Australia. In those two colonies they had grey and black sulphurets of copper, but in the Peak Downs they had black oxide of copper and malleable copper in quartz. One of the most interesting sights when he visited this mine was the raising of copper in quartz. This only required one treatment in the furnace before being refined."

Arriving on 22 November, Elizabeth began settling into the manager's cottage, one in a row of cottages, while Josiah arranged with the managers to view what progress had been taken since his previous visit. Before his arrival, a fire had broken out on 29 October near Thomas's and Joubert's shafts and was still raging after three days, with intense flames and smoke. Eventually it burnt out, the supporting beams and shaft having collapsed, extinguishing the fire. The cause was unknown, but generally believed to be started by a candle being left near dry wooden beams. Captain Terrell, the mining manager whom Josiah had met previously, had resigned on 5 November and so he was introduced to Captain Hill, the new mine manager. Hill had instructed that a new shaft be sunk just north of Nichols shaft on the western portion of the property. At the 14 fathoms level, a lode was discovered, and driving east 6 fathoms following the lode, the quality improved. What was even more encouraging was the fact that the remarkably soft ground would significantly reduce mining costs. Fully 4 ft wide, it was well defined, displaying rich green and blue crystallised carbonates of good quality. Having again inspected the controversial Josiah shaft and Mitchell's shaft, Josiah decided that no further exploration or mining should take place. Similarly, as the ore had petered out in the Jerusalem and Old Engine shaft, he directed the men to other projects. By connecting a winze to the drive of Jobour's shaft, a lode 3 to 4 ft wide yielding about four tons of ore of 18 percent copper was discovered. He was hopeful this lode would continue.

On the eastern side of the mine site, Andrew's vertical shaft had been sunk to 19 fathoms and proved payable at a very short depth, but the deeper area containing ores were small and irregularly defined. He had instructed that the shafts be sunk from 27 to 40 fathoms east of Andrews shaft, one a vertical shaft and two underlay shafts to depths of 6 to 11 fathoms. He was confident that extending the drive would provide a better exit for hauling the ore and removing water that was now beginning to impede progress. Around Andrew's shaft, further development was planned given results of testing had shown a strong lode outcrop. After completing inspection of the shafts, Josiah recommended that a deep shaft be sunk at a short distance south east of the old engine shaft to intersect the lode at 70 fathoms, the cost estimated from £6000 to £8000 and would take two years to complete. Before undertaking a work of this magnitude, the directors needed to ascertain what reasonable hopes there were of ensuring "… sufficient quantity of payable ores to defray its cost, and at the same time cover current expenses".

With the general manager Woodhouse, Josiah also spent some time checking that the mine was well ventilated and the shafts were well-constructed as well as inspecting the smelting works and all the company's plant. This included machinery and hauling gear, four engines, two to pump and wind, one to pump water for dressing purposes, the other awaiting assembly. Having completed the inspection and satisfied himself that all was running well, and that he had provided direction for continuing excavation of the various shafts, winzes and adits, he began preparing his report for the board for the six months to December 1873. He began:

> *Sir,— Having taken the management of the mining department so recently as the 24th November last, and having now to report upon its development for five months preceding the 30th of that month, you will, I trust, excuse me should I fail to touch upon all topics connected with the department. My chief aim will be to lay before the board: the present condition of the mine and as early future prospects.*

His report was extensive as it was comprehensive, detailing each of the shafts and outlining the interaction between them, the depths and quality of ore at various levels. With the increase in production, Captain Thomas decided to start the refinery once he had procured the necessary number of competent staff/hands.

During the year, another extensive property was acquired, the Western Peak Downs mine, situated about seven miles west of Copperfield. This mine had been taken up some time before by the Western Peak Downs Copper Mining Company, Limited, (WPDCMC) which, having been floated on inadequate capital, had now shut down. Having only partially opened out various pits and shafts, the mine was disposed of for a few hundred pounds to the older proprietary PDCMC, which hoped to supplement its supply of ores via a tramway connecting the two mines. This hope was, to a great extent, frustrated by the comparatively small value of ore and the prohibitive cost of a tramway to connect the two mines.

After undertaking a complete review of the labour force, Josiah dismissed 131 unnecessary men, thus making a saving of over £12,000. Since taking over the mining operations, he had planned and designed more than 900 fathoms of new works, including new shafts, expanding existing drives and implementing adits as required. This had been achieved with a work force of just over 260, managed by four supervisors or officers, a considerable reduction from over 400 six months earlier. Also, as a large share of stoping was done on day work by revamping the workforce nearly all underground work was let on contract or on tribute with considerable savings. One of those to leave was smelting manager, Captain Thomas, who, perhaps after hearing how much Josiah was being paid, requested a pay increase.

During the first six months of 1874, over 5000 tons of ore was extracted for smelting and 107 tons received from other mines in the district. Combined with the ore waiting for smelting, more than 7000 tons was smelted with a yield of 16 percent copper, somewhat less than the 20 percent earlier. That produced almost 750 tons of refined and rough copper of which almost a quarter was sold in Sydney. However, because of the conditions the carters were facing,

300 tons were on the road between the mine and port and over 200 tons were sitting at the mine site awaiting collection. This problem would be a continual one severely impacting the company's cash flow. In his 28 August 1874 report to the board and shareholders, Captain Holman wrote:

> *It will be observed large reductions have been made in the mine force since January. They were then too overcrowded and necessitated carrying on more dead work than desirable but nearly all the miners being under engagements they should not be disposed of at will. However, those that were not indebted to the company and applied for leave had it granted. Several were paid off as their engagements expired and several absconded. The present forces are now more easily and better placed and being in smaller parties, they work more beneficially for the company and themselves.*
>
> *In conclusion I estimate the returns of copper ores to be sampled in the ensuring six months at from 800 to 900 tons monthly 14 per cent copper by employing about the present forces.*

Being also responsible for the West Peak Downs company, he visited the site in September 1874, and after a careful examination of the mine site and operation, determined:

> *The result of developing over 200 fathoms of exploratory work in this lode has shown a decidedly non-paying result, and I do not see any better prospect in the future. Instead it began with traces of carbonates, followed at the twelve fathoms' level with yellow sulphurets of copper that have deteriorated at deeper levels, and is now wanting in promise to become payable in the early future. Therefore, my opinion is decidedly against advising you to resume explorations on this lode, as all moneys invested therein will, in all probability be totally lost, the returns of ores adding merely to keep the works protracted, but eventually returning nothing to the investors.*

At an adjourned special meeting of the Western Peak Downs Copper Mining Company held on 30 September at 5 Spring Street, Sydney, reports of the directors and Captain Holman were read and a vote was unanimously carried: "That the company be forthwith dissolved and wound-up, and that the directors be instructed to take the necessary steps for carrying: out such dissolution in terms of the deed of settlement."

Meanwhile under Josiah's direction, work had continued at a fast pace; at the western section just below the 29 fathom level, a shaft situated north-east of the Nicholas shaft had produced 95 tons of ore, 30 of which produced 24 percent copper. At Joubert's, Thomas and Engine shafts the stoping had been gradually discontinued and a large portion of deserted and older mines were thrown open to tributers who vigorously worked the ground. Many who decided to overhaul the previously worked lodes were rewarded handsomely. The Mitchell shaft had been extended 18 fathoms west and 13 fathoms east following a lode of payable black ore averaging nearly 17 percent copper. This encouraged the drive to be enlarged and allowed for the completion of the double line of railway track. Sinking of Hancock shaft continued below the 21 fathom level and Pearce's shaft was producing ore of around 10 percent copper at the 15 fathom level. By the end of 1874, with the arrival of the hauling gear ordered from Sydney, 3330 tons of ore had been raised on tut-work and 3974 ton of ore on tribute producing over 1000 tons of pure copper, the average assay 14.24 percent; value £80,000 at the prevailing rate of £80 per ton. This exceeded estimates considerably, partly due to increasing the mine work force, now over 400 to excavate the rich ore producing shafts in the old mine and the eastern sections. A reporter visiting in 1875 wrote:

> The directors have secured the services of a first-class mining captain — Mr. Josiah Holman — acknowledged to be one of the most experienced men of his class in Australia.
>
> *Australian Town and Country Journal,* Feb 1875

However, Captain Holman's greatest concern was still the smelter operations, hindered by the ever-increasing cost and dwindling supply of fuel as well as ongoing delays being experienced in getting the copper to port. There were 24 furnaces, of which 19 were reducing furnaces, three calcining furnaces and two refining furnaces all requiring vast quantities of iron bark and box-wood. As no coal was available, 5000 tons of firewood per month would barely meet the requirements of the smelters. Over the years, the nearby forests for miles around had been stripped bare creating a scarcity which necessitated that the teams go greater distances for timber.

Peak Downs QLD, painting by Conrad Martens – note the number of chimney stacks

After 12 months, Josiah and Elizabeth had settled into the community well, although Elizabeth found the heat and humidity during the summer almost unbearable; it was beginning to have a detrimental effect on her overall health. Just before Christmas, they joined many of the families and enjoyed watching a procession led by a brass band which terminated at a nearby field. Here they were entertained by foot races, by age, high jump events and a billy goat race much to the amusement of all. This was followed by an afternoon tea, music and recitals. On New Year's Day, the children were treated to tea and cake on a humid, sweltering 110-degree summer's day.

Robert Morehead had kept in touch with Josiah and continued to convince the SAMC board that the Cadia mine offered great potential, especially if additional capital could be used to extend the mine operation. Also, prices of copper had begun slowly rising from lows of around £65. In March 1875, he wrote asking Josiah if he could outline what workings would be required and the cost to extend the mine's capacity. On 20 April, Josiah replied with a full and detailed report, including three sets of plans, outlining what would be required stating:

> ... the property should be prosecuted for gold and copper simultaneously, because the former metal, and greater or lesser degree, seems to have been generally found in the copper or that exists in the properties and that the abundant water power on the property should be fully utilised in conducting such operations ... see plans numbers 4, 5, 6.

He further strongly advised that the company, in the circumstances of the case, should take the working of the property into their own hands, rather than allow third parties to lease workers on a royalty.

Later, Josiah was again contacted by Robert Morehead and asked if he could visit the nearby the Dee copper mine purchased in August 1874 for £20,000 by SAMC. Managed by 45-year-old John Henry Holman, of Melbourne, the mine sat on over 1500 acres boarded by the River Dee about 15 miles south of Westwood. Travelling almost 200 miles to Westwood in late October, he spent almost a week inspecting the Dee copper mine. In his report of 3 September, he recommended extending a number of the exploratory shafts, especially those highly productive yielding 20 percent copper. He noted that there was ample fuel close by and cartage costs to the port of Rockhampton were a very reasonable 4s per ton.

Ever so slowly, Peak Downs began to emerge from its embarrassed financial position, each succeeding six months showing incremental improvements. By the end of 1875, the whole of the debt had been wiped out and the estimated assets of the company showed a surplus of £52,000. For the previous six months, 4104 tons of ore had been mined and 4626 smelted at an average rate of 13.57 percent producing 546 tons of refined copper. Of that, 436 tons was shipped to England and 110 tons sold in Sydney and at the mine or in transit were over 1000 tons of refined copper. Compare that with Burra Burra, now an open-cut operation excavating tons of "orey stuff" (rock) which, because of difficulties experimenting dressing and processing the "orey stuff" produced similar results, but at a significantly higher cost. In fact, the open cut method was abandoned in 1876 and 90 or so tributers and tut men were engaged. But that was short lived when the mine closed the following year. Peak Downs was now the largest mining venture operating in Australia.

In Josiah's report of the 31 January 1876, he wrote:

> ... have not made any important discoveries during the half year closing 31st ultimo, yet notwithstanding the production of ores have again shown a notable decrease, it has been attended with profitable results. The explorations have not been carried out upon a large scale; the chief of which has been sinking and testing the lode in depth.

He further stated:

> The production of ores in the ensuing six months will continue to decrease unless payable discoveries are made in the deep explorations. The shallow explorations apparently afford but little hope of sustaining the returns and the reserves generally show signs of rapid exhaustion.

Of the Western Peak Downs Mine, he wrote:

> This property has been developed by testing the reserves of ores in the roof at the 22 and 12 fathom levels by Tributers. The result has been non-paying in general and the tributers have abandoned it, whilst every encouragement was held out by giving high rates of tribute.

Over the next 12 months, with no new discoveries, the mining payroll had also decreased, the number of miners employed now averaging 173 made up of tributers 130, tut-work 43 men as well as eight boys plus trammers, fillers, surface labourers, mechanics, engine-drivers etc. By now Josiah had lost faith in the mine's longevity and with only enough ore to keep seven furnaces in operation and in his mid-50s, he decided he had enough of the rigours of mining in remote Central Queensland. Also, the wet hot summers were affecting Elizabeth's health; she did not wish to endure another year and was keen to return to the cooler dry Cadia climate and the children. Or was it the fact that Morehead, keeping in close contact with Josiah and having secured additional funding for the Cadia operation induced him to return? After providing notice, he and Elizabeth packed their belongings, bade farewell to their new friends and left Peak Downs by coach in April 1876.

Having secured Josiah to manage the Cadia mine, Robert Morehead asked if he could stop by the Dee copper mine on his return home. Josiah spent a few days inspecting the operation with John Henry and was impressed with the progress made with the furnace to convert the ore into regulus (mass of metal, in an impure state); it was expected these copper ingots would be shipped out within the next few months. Returning to Westwood, they caught the train, a 30 mile trip to Rockhampton and boarded the 415 ton *Boomerang* on the 19 April 1876, arriving in Sydney on a very quiet Sunday 23 April. They were greeted at the docks by an extremely excited young Annie who had been given permission to leave the boarding school at Petersham to meet her parents. Checking into Tattersalls hotel, they spent the day with her before walking her to Central Station where she caught a train to Petersham. Elizabeth spent the next few days shopping and catching up with friends, Josiah meeting Robert Morehead and discussing the Dee and Cadia mines. It was agreed Josiah would continue working the Cadia mine as he had before, paying a royalty of one-tenth part of all copper ores and one twentieth part of all gold smelted. So as not to incur further expense for the company, he was to make use of tributers as and when required. He was to report on a six-monthly basis in time for the half yearly board meetings and should he believe further capital needed to be incurred, he should include that request for the board's consideration. Returning to Cadia, Josiah and Elizabeth caught a train to Bathurst, the extended line from Kelso having been completed earlier in the month. Josiah smiled as he thought how far things had come in this country since he first arrived when he had to disembark at South Creek to catch a coach. The enthusiastic greeting by the children made them realise just how much they had missed their children and Cadia.

Given a discouraging report at the end of 1875, and the fact that Josiah would leave a large gap in the management of the Peak Downs mine, the directors arranged for Mr Samuel Higgs, F.R.G.S. (Fellow of the Royal Geographical Society) and general superintendent of the Wallaroo Copper mine in South Australia to visit and inspect and report on the mining operation. Copper continued to leave the mine in drays heading for the coast where shipments were now being taken from Rockhampton directly to London. During the six months to 30 June 1876, less than 3000 tons of ore was raised and over 4000 tons of ore was smelted to be sold to a declining market, copper prices having earlier risen, now languishing around £80 per ton. Higgs had made suggestions regarding additional shafts that should be sunk and both he and Josiah were asked to provide an estimate of cost for such works. At the half yearly general meeting held on 31 July 1876, a resolution to call a special meeting to dissolve the company was defeated by a majority of five. No dividend was declared.

Almost 12 months to the day since Josiah had left Peak Downs, a notice of a special general meeting to be held on 25 April 1877 was published. The purpose of the meeting was to vote on the following resolution: "That in view of the present prospects of the mine it is advisable to dissolve the Company ..."

At a meeting on 25 April, the resolution was carried, 297 to 87 against. On 20 June, tenders were called for the purchase of the property, 800 acres complete with a mining and smelting operation described as being in efficient working order, final tender date 12 September. In October 1877, a prospectus was issued for the Peak Downs New Copper Mining Company. This new entity made an offer of £5000, but was rejected by the Peak Downs directors. Clearly the directors of the PDCMC did not feel this was enough, for on 20 November, Mort & Co.,

Auctioneers advertised a public auction for the mine and plant of PDCMC to be held in their sale room in Phillip Street, Circular Quay on 18 December. Given the history of the mine, which one scribe described as "blunders and plunders", it is not surprising there was little interest and the final knockdown price of £3000 was paid by Messrs Thomson and Cashion on behalf of other parties at Rockhampton and Copperfield. Over its lifetime, a total of over 11,400 tons of copper was sold realising almost £1 million, of which dividends paid to the shareholders equated to almost 20 per cent per annum.

Christoe always had a strong belief in the mine and in January 1878 was appointed acting general manager of the Peak Downs mine which would be let on tribute to existing miners. Mounting capital outlay closed the mine almost a year later. In March 1879, a meeting of shareholders of the new Peak Downs New Copper Mining Company voted to accept the only tender offered to continue mining operations. Christoe believed he could operate the mine profitability and with the help of many of the miners who had not left the area, he had some initial success. At one point, more than 200 men were engaged in various capacities working the mine, but this was short lived. Christoe was declared insolvent and when a catastrophic fire burnt down his office, it destroyed his experimental developing laboratory. Much like Josiah, Christoe never moved far from the mine he believed in and stayed in and around Central Queensland before he died at Mackay in Queensland in October 1918.

Retirement, Pastoralist

1876–1893

At the Scottish Australian Mining Company's half yearly meeting of shareholders held on Monday 29 November 1875 at the London Tavern, Robert Morehead reported that iron ore of excellent quality had been discovered at the Cadia property. He further noted that continuation of the great Western Railway from Sydney was expected to be completed to Orange towards the end of 1876. On 3 September, he had written to the board:

> I continue desirous that this company should itself undertake the working of the property for gold and copper. On putting the question to-day to Capt. Josiah Holman, he tells me that an amount quite within £10,000, possibly not more than half that sum will suffice, he believes, to open it out in a satisfactory manner. I hope, therefore soon to receive an intimation of approval of working Cadia on the system set forth in his letter of April 20 last.

Reaching Cadia, they were greeted warmly by John Trathen the superintendent of the SAMC Estate. Trathen resided in a fine brick home which stood in the middle of a lawn of English grasses, enclosed within a large tall fence. They had missed the trees changing as the seasons rolled one onto the next, just as it was in Cornwall. It was so nice to be home in their comfortable property, with its splendid hand-made brick house, lovely garden and orchard of fruit trees including mulberry and peppercorn spread across the lawned area. They had missed the children and were eagerly greeted by Joe, Elizabeth noting he had lost a little weight. Charles had left early to inspect one of the fences after hearing from a neighbour that some of his stock had escaped. Emily had written to say all was well with their new son Henry Ernest and they intended to return to Cadia from Oban sometime next year or the year after.

Village of Cadia

The village of Cadia had shrunk even further than when they had left almost three years earlier. Josiah spent some time talking with the few miners still working various shafts and decided he would explore further the White Engine mine shaft to the west. The east Cadia area had been worked on tribute while he was away, a team of men having raised over seven tons of ore which produced almost 10 ounces of gold. Gold nuggets were also found in the alluvial areas one weighing almost 40 ounces. He spent some weeks working systematically, following each of the copper and gold-bearing lodes in the hope they would lead to richer, more substantial bearing yields. He was able to engage 14 tributers, some assisting with the deep mining of shafts, others with smelting experience preparing one of the furnaces for operation. Hearing that Captain Holman had returned to Cadia and was pushing forward with the mine operation, within a few months 35 tributers were working in various locations of the mine. Over the next two years they were able to extract over 1000 tons of ore averaging 11 per cent copper.

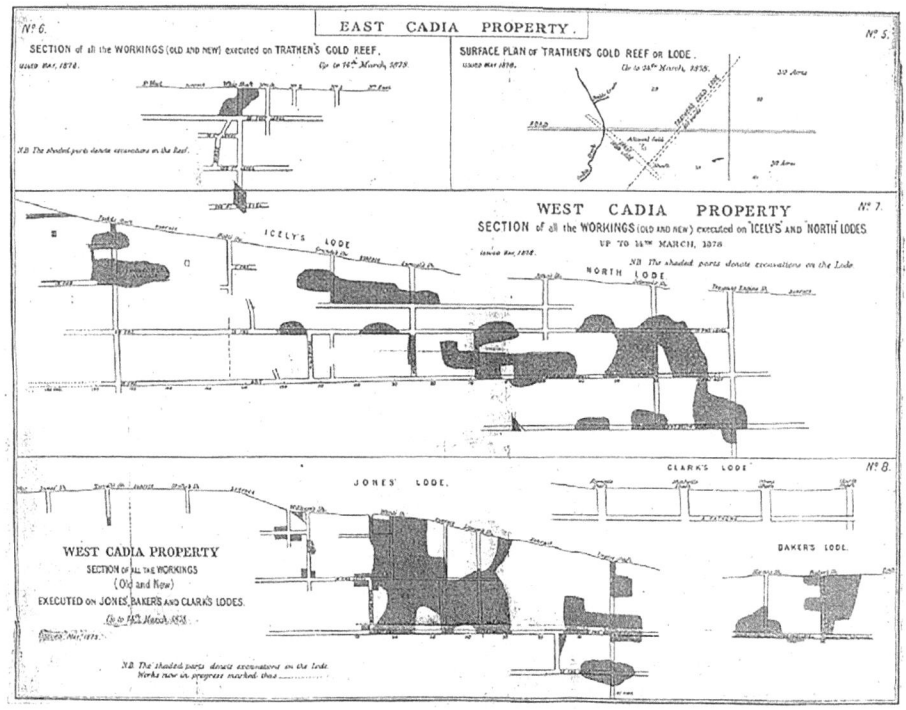

Cadia mine shafts – circa March 1878

After considerable agitation for reform, the Robertson Land Acts (Crown Land Alienation Act) of 1861 was introduced and provided a mechanism whereby smallholders could purchase Crown land. Gustavus Richard Glasson was one of the settlers to take up portions early on and quickly accumulate a large holding, others would follow. Dr Mathew Blood took up 100 acres on 17 July 1862 as did Richard Stanley Wilson taking up a number of portions and by early 1876, he had taken up nearly 1000 acres comprising Portions, 1, 2, 3, 4, 5, 20, 23, 24 of the Parish of Blake. On 22 July 1876, Josiah and Thomas Honey applied for lots 6, 7, 8, 9, 10, of the County of Hardinge, Parish of Blake, each lot comprising approximately 20 acres. When the opportunity arose to purchase some of the portions from Richard Wilson in November 1876, Josiah bought most of the portions, comprising around 830 acres, which he named Boxland Park, which was added to his pastoral holdings.

On 1 December 1877, the 576-ton steamer *SS Rotorus* arrived in Sydney Harbour, Mrs Elizabeth Robson on board having left Auckland on 20 November. Leaving

Sydney, she travelled by train to Orange where she was met by an elated family. It had been over 12 years since Josiah had seen his eldest daughter, it would be a joyous Christmas, the first time in 15 years so many of the family had been together. Emily Louisa, expecting their fourth child, and William also returned to Cadia from Oban. There were in all five children and three grandchildren present for Christmas. On 23 January 1878, another celebration ensued, the marriage of Josiah Jr (Joe) Holman, now aged 24 to Emma Burfitt, eldest daughter of James Ingram Burfitt, a local farmer. The wedding was held at the Church of England Holy Trinity Church, Orange, officiated by Rev F. B. Boyce and afterwards moved to the Burfitt home in Cadia. Elizabeth stayed on for three months, updating them on life in Auckland and returned on the *SS Rotorus*, departing Sydney Harbour on 10 April 1878.

For the Scottish Australian Mining Company, Cadia, one of a number of mineral holdings, continued to disappoint. At the half-yearly meeting held in early February 1878 at the City Terminus Hotel, Cannon Street, London, it was reported that:

> With regard to the Cadia properties, the position of matters was much the same as in Queensland, but unfortunately the ore found there was not of a high quality; however, plenty of copper had been taken from there in former times. The gold reef had been at present extremely disappointing, for they had quite expected at one time to get an ounce per ton, but at present it was very poor indeed. One of two things must have occurred; either they were on a very poor part of the reef, or there was a mistake in extracting. They had spent £10,000 on this property, which consisted of 3000 acres.

Later the directors agreed that:

> The Cadia property had been worked for copper, and at this property they used better appliances for the production of metal than they had at the Queensland mine, and the directors recommended that for the present operations at Cadia should be continued.

In October 30 1878, the directors reported:

> Working at the Dee site has been stopped, and it appeared that with regard in the Cadia it was suggested that a large amount should be expended in improving its value, but we would not propose that course with the present depressed price of copper. They must either let it lie idle, work it themselves, or get someone else to work it for them.

Josiah and his small team were providing a small revenue stream for the SAMC but had yet to make any substantial discoveries. Copper prices had languished around the low £70s per ton for some years and continuing their decline, by 1879 were trading around £60 per ton. SAMC were in somewhat of a quandary as to what to do with the Cadia property, having observed how little unproductive mines would fetch on the open market. As an investment, its returns were far from adequate and with low copper prices, the property as a mine had little value. In considering the Good Hope Copper property, nearby at Yass — a property which some day might be sold — the directors proposed writing off about £4000 of the capital which had been expended in mining operations.

Previously, in 1866, Cadia residents and miners had successfully petitioned for a Common to be declared around the mine to reserve these resources for general use. A consequence of the Common was that it restricted conditional purchases in the area until 1879 when the Common around the mine was revoked. This change allowed Josiah to add to his property holdings and on 17 April 1879, he applied for Portion 91, 100 acres and on 5 June applied for Portion 92, 40 acres, on 15 April 1880 applied for Portion 95, 80 acres, on 30 September for Portion 96, 40 acres and on 11 November Portion 97, 97 acres. He continued to increase his property portfolio, taking on conditional purchases on parcels of land as they were offered as well as obtaining transfers from other owners. He had also negotiated to use the land owned by SAMC for grazing sheep, paying 68 shillings a year.

Later years of Cadia mining

At the half-yearly meeting of the Scottish Australian Mining Company held on 17 November 1880, at the Cannon Street Hotel, London, the issue of the Cadia copper property was again raised. The board believed it was fundamentally a good property, comprising 3000 to 4000 acres of freehold land currently let on royalty and suitable for other purposes than just mining. Lately, copper had been sent from Cadia and with copper prices firming over the past six months and now trading at more than £80 per ton, some had been sold. The board did not wish to part with the rest, hoping prices would rise further:

"With respect to the Cadia property, that still stood in the accounts at much the same figure, Captain Holman was yearly tenant there and the company was under no expense in regard to it. The property was known to be auriferous; specimens of rich quartz having been received from it a few years ago and Captain Holman was now giving increased attention to it with the hope that a rich reef would be found. The board encouraged men to search the property under the auspices of Captain Holman at their own expense by offering them a large share of anything they might discover."

Having achieved only moderate success, tributers slowly drifted away but Josiah, as lessee, continued exploring various areas of the mine site that had yet to be fully tested. He was able to mine, crush and smelter a few hundred tons of copper ore which was sold in Sydney. During the six months to the end of 1879, he and the remaining tributers received over £500 from copper sales. In or around August 1880, while sluicing dirt and gravel on the hillside behind Roberts' old hotel, he was able to flush out five fine specimens of gold, one nugget weighing 11 ounces, the five giving an aggregate of 20 ounces of fine gold. From the SAMC Boards position, this only confirmed "… the opinion long entertained by the board that the property is auriferous in its character, and they believed that the reef whence, in past ages, these nuggets of gold, and others previously obtained, had been washed down, could not be very far distant from the spot where they were found; and with the object of discovering it, instructions had been given to Mr Morehead, the manager, to procure the services of a competent independent gold miner from Victoria to examine and report upon the property, and to advise whether and in what parts it might be worked for gold profitably."

Chilcott Street, Cadia

Following a directive from the board, in early 1881 Morehead contacted James Munday, a well-known gold mining engineer in Victoria, asking if he would be available to visit the mine. Arriving in Sydney on March 14 on the 1200-ton steamer *To Anan*, Munday made his way to the SAMC offices in O'Connell Street. He spent most of the late afternoon and evening with

Morehead going over in more detail the assignment and the following day, caught the train to Orange. There he met Josiah and on the way to Cadia asked him about his gold discoveries over the past few years. Munday was able to stay at the Bon Accord Hotel and spent the next few weeks prospecting and fossicking around the mine site, examining the shafts and paying particular attention to where gold discoveries had been previously made. He also collected a number of samples of various rock formations and gold, if he was lucky, for further examination. Returning to Melbourne, he wrote his report and sent it to Morehead in Sydney. He determined that:

> "Although I found, in the course of my examination, no particular part of the properties to be very rich in gold, yet that metal proved to be very generally disseminated, more or less, throughout them at and near to the surface. From the alluvium at shallow depths and the reefs on the properties, gold to the extent of 329 ozs. altogether has from time to time been raised; in one instance (in 1873) a nugget of upwards of 39 ozs, and in November 1880 several nuggets, varying from 1 oz. up to 12 ozs., were found.
>
> The deepest working on the reef does not exceed 163 ft High reefs, more or less showing gold, have already been discovered on the properties. Past operations on the properties have proved that they were very rich for copper at and near the surface, ore having from time to time been raised from them which has yielded, 976 tons of fine copper.
>
> The works, though numerous and extensive, have all been near the surface, the deepest level not exceeding 25 fms. from the surface, except a portion of an adit into the sloping hill, the end of which may be said to be 40 fms, from the surface. The ore raised has yielded 9, 12, 14, 17, and a considerable portion of it 20 per cent, of copper."

Hardly conclusive, but it did add weight to the boards belief, long entertained, that the property is auriferous in its character, and others who, having inspected the gold nuggets were of the opinion a rich gold reef "… could not be very far distant from the spot where they were found."

The discovery of copper and quality iron at Blayney some 15 miles to the south in early 1882 caught Josiah's attention, as more recently the ore being excavated at Cadia contained both copper and iron as well as occasional gold. After inspecting the Blayney mine's huge copper deposit, he wrote to Morehead noting that the stratum formation was very similar to that of Cadia.

During 1882, the total quantity of gold, including one large nugget weighing 76½ ounces, amounted to 297 ounces 16 dwts 15 grm of which Josiah and the tributers shared almost £1000. Reporting to the board in January 1883, Josiah wrote:

Along the outcrop of this vein, for the short distance was explored and some very nice loose gold was obtained in the Ironstone when crushed in a mortar yielded fair prospects of gold. I am about to put two men to take out a few tons of this vein and hope to get it crushed during next month with the object of ascertaining if it contains gold in paying quantity.

Buoyed by this success, Josiah arranged to increase the supply of water for sluicing by deepening the tail races to the alluvial areas. Also, the water-race to the stamping plant was repaired and rectified and SAIC procured a waterwheel to connect to the stamping plant which would be used instead of steam power to drive the wheel when there was sufficient water. Josiah believed it would ... *make the crushing charges lower and thereby encourage miners to test the reefs on the property.* He further wrote on 20 January 1883 ... *I have already sub-let to two men the north-eastern portion of the sluice claim and these are now bringing up a deep tail race for following a run of gold in this portion of the claim. I expect shortly to sublet the western portion of the claim.*

Josiah's lease was extended a further 12 months from January 1883.

William Blood and Emily having returned to Cadia in 1877 had another daughter, Dora Ann born on 28 September 1878. It was unfortunate that the baby died after four months. In May 1883, William Blood was appointed postmaster, but sadly on 4 October, his wife Emily Louisa died while giving birth to their fifth child, Susan Louisa. Emily was buried at the cemetery in Cadia. William later remarried and died at Cadia in March 1905.

It was around the end of 1882 that the Morgan brothers discovered auriferous deposits of significant magnitude and staked their claim on 640 acres located towards the end of Dee River valley, 35 miles south of Rockhampton. Rising almost 500 feet, the well-defined hill stood out distinctly from the surrounding undulating ridges and rolling plains. It is not surprising that what followed was another stampeding gold fever, just like those that had swept the country for the past 40 years. Replicated time and time again, prospectors searched in all directions near the "mother lode" in the hope of staking a claim across acres of precious metal that would set them and their families up for life.

As reports of this extraordinary find filtered through, Morehead suddenly became more interested, as it was not too distant from the Dee copper mine. He contacted Josiah and asked if he could travel to Central Queensland and inspect the newly discovered Mount Morgan gold mine. On arrival at the site, Josiah was surprised at how much gold had been uncovered so close to the surface, easily allowing for the development of an open-cut quarry. There were more than 100 men toiling in and around the hillside, carting ore from shafts and tipping the ore into a deep wide wooden race that had been constructed over 200 ft long. The slope of the race moved the stone down the hillside to a level area where it was then carted a few hundred yards to the two batteries. The stone was then crushed by the two batteries, one using 10 stampers the another 15, under the protection of a huge galvanised iron roof. Water, always an issue — too little or too much — was conveyed through a series of pipes from a large waterhole nearby. Examining the mine works more closely, Josiah noted the strata formation was very similar to that of Cadia. However, as to the reefs that might continue towards the nearby Dee mine, he would need heavy excavation equipment to determine that.

In late 1884, Government Geologist Robert Logan Jack visited the area south of Rockhampton and tabled a report in the Queensland Parliament on 21 November 1884. He was of the opinion that the discovery was "… one of the most important events in the history of the mining industry."

At the SAMC half-yearly general meeting held at the Cannon Street Hotel, on 21 April 1885, it was reported that:

"Captain Holman paid a visit recently on behalf of the company to the famous Mount Morgan Gold Mine, with a view to ascertaining whether the work being done there will be likely to throw any light upon the question whether this company's copper property near Rockhampton, which lies within a few miles of Mount Morgan, might be expected to be found to be auriferous. As regards this point no intelligence of a decided kind, one way or the other, resulted from Captain Holman's visit."

Cadia main street, Chilcott Street with GPO (Post Office) on RHS

With the number of tributers having continually dwindled, Josiah approached the SMAC board suggesting the current one-year lease be extended to three years. Such tenure would provide greater stability and attract more miners to the site, he argued. The board was unmoved, but still in a quandary as to what to do with the Cadia property. In April 1884, the SAMC board reported that ... the quality of gold obtained since the previous August (1883) had been 82 ounces 17dwts 17g and during the year mainly from alluvium 118 ounces, 4 dwts and 1g realising £411 11s 11d.

A severe drought impacted the sluicing operation but construction work continued on the eastern tail race, extending it to 35 fathoms to intersect with a deep run of newly discovered alluvial gold. On the western portion, the tail race had been extended to 45 fathoms. He had engaged two men on the Iron Duke claim who had excavated almost 50 tons of ore with a payable yield. To smelter the heaps of ore which were piling up, he directed a number of men to rebuild the copper furnace.

Cadia mine site showing the tail race.

While the mining operations took up much of his time and was remunerative, work still continued developing the grazing property. By 1884, he had accumulated around 5000 acres to the north-west of Cadia village, comprising Boxland Park and Tunbridge Wells supporting almost 4000 sheep. He recognised such enterprises demanded a greater level of capital than most possessed and that success depended on building up a sufficient landholding to make farming and grazing viable. It was only these larger and thus more economically viable properties who were able to productively and profitably run sheep. But even so, it was still necessary to have family support during hard times. There was much celebration in the Holman household when Annie, the youngest sibling aged 25, was married on 27 February 1884 to

Fredrick Parish, a clerk working for Dalton Bros' merchant store in Orange. They lived in Orange for another decade.

Robert Morehead, who had established the SAIC in 1840 and was now 72 years old and in poor health, resigned in December 1884. Recognition of his long and faithful service was rewarded by the board with free rental on the property in which he resided in Phillip Street as well as a lifetime pension of £1000 per annum. It was unfortunate he did not live long to enjoy these generous benefits as he passed away on 9 January 1885. He was survived by a son and two daughters. Since arriving in 1841, Morehead had been involved in copper-mining in South Australia, NSW and Queensland, coal mining at Newcastle and large-scale pastoral ventures in Queensland, particularly at Bowen Downs and the Gulf country where SAIC was the first to recognise the great opportunities northern Queensland offered. He had built and established a vast empire which included the largest pastoral station in Australia (Bowen Downs), a significant and extensive property portfolio in the city of Sydney. He also developed the Hunter Valley coalfields into the most productive in NSW. The company's capital increase was truly remarkable, from £30,000 in 1840 to almost £1 million in 1885. It was a sad day for Josiah, hearing the news of the passing of a good friend with whom he and his family had built an association over the past 22 years.

From the various trials, gold in small quantities continued to be found during sluicing or when crushing stone from the reefs of Iron Duke, Owens, Trathens and East Cadia. Nuggets of various sizes were also occasionally being discovered scattered about the property. Josiah wrote:

> *From the finding of such nuggets from time to time and the fact that gold has been proved to exist in semi-different places on the property, although not yet in large quantity, it would appear that there must be a source not far removed from whence the gold so far met with has terminated and which, if it could be discovered, might prove to be of much importance to the company.*

In mid-February 1885, while Josiah was fossicking and sluicing in an area beyond Roberts Hotel, a little further from where he had, a few years earlier, uncovered a number of nuggets, his pick struck a large rock. Using the pick to gently dislodge it from the gravel, he discovered it was almost 10 inches long, and once cleaned, its yellow hue shone brightly. He knew immediately it was gold and would later determine it to be pure gold — a nugget weighing around two pounds (32 ounces). He could hardly contain himself as he ran home and announced the news; celebrations would surely follow. On the 27 March, after sending invitations to friends and family, a party was held at his home Rosemont. People came from far and wide on horseback, carriage and dray, arriving in the late afternoon, the celebration carried on well into the night, with singing around the piano and dancing until dawn. It was reported that Josiah's wife was dressed in a stylish black silk dress, daughter Annie in black Grenadier and Scarlet ribbons, Miss Ada Stevens in black and silver, Mrs Margaret Blood in a gold satin dress, Miss Emily Blood in a pretty blue-and-white dress, Mrs Emma Holman in cream silk, Miss Stevens, Ada's sister, in black and silver, Miss Mabel Holman, Josiah Jnr's daughter, in blue silk and Miss Connie Webb in grey and pink.

But that was only the beginning of much excitement in the Holman household as they soon began preparation for the wedding of their youngest son Charles William. On 26 May 1885, he married Ada Mills Stevens the daughter of school teacher Walter Stevens at the English Church, Canobolas, officiated by the Rev. T. D. Dunston. More celebrations would follow when, in late 1886, Charles discovered a very large nugget weighing 70 ounces[*]

At a location near Dry Creek, Millamurra (Millah Murrah) 10 miles south of Sofala and 22 miles north of Bathurst, the proprietors of Mount Rosette Gold Mine wished to float a company comprising five acres of Auriferous Land, together with a crushing plant in good working order, valued at £800,

---

[*] The Powerhouse Museum in Sydney has in its collection a model cast, purchased from Mrs A G Goodman with a number of other plaster models of gold nuggets and minerals. https://collection.powerhouse.com.au/object/2232

for the systematic working of gold mining and crushing. Josiah was contacted in early March 1888 and asked to survey the mine, plant and operation and provide a report for the prospectus. Having travelled 50 miles via Bathurst, he arrived on 19 March and the next morning began his assessment. Checking the shafts and stopes, he could see the veins of quartz contained patches of gold and later noted the nearby creek had been worked extensively for gold. He then drew a sketch of the mine operation with notes on what proposed new works he would suggest to maximise the mining operation returns. This included crosscut tunnels and methods of hauling the quartz using buckets and a series of ropes to the battery floor. The mining plant comprised a boiler, 8 hp engine, a battery of six stamps with the option of adding six additional stampers, all under a substantial galvanised roof building. From the small hillock, he could see there was plenty of quality timber close by for fuel and building supports for the shafts. In conclusion, he wrote: *I can recommend with confidence the Mine as a good investment.*

Josiah had spent almost a decade searching for the source of the alluvial gold he and the dozen or so tributers had been sluicing and mining. They had followed numerous reefs within the Cadia property around Iron Duke, but rich fortunes had not been made. For the miners, it had barely kept enough food on the table and by 1892 only a handful of tributers were attempting through sheer hard work to generate a living at Cadia. At the half yearly meeting of SAMC held in London on 29 April 1892, the board reported "... the mining operations had practically ceased". Only 9oz 18dwt of alluvial gold had been raised by tributers and 12 tons of copper ore at 20 percent was smelted, realising £43 15s per ton, a far cry from the £100 decades earlier. Meanwhile, Josiah, continued building his property portfolio, purchasing Lots 127, 93, 124 from Stedman, Ferris and Patterson in 1891.

In 1892, Elizabeth Robson left New Zealand, perhaps compelled by her father's declining health, which worsened during 1893. Confined to bed at the onset of winter 1893, Josiah was treated for fever by Drs Codrington and Kelty but died on18 September 1893. Cause of death was reportedly fever and ague contracted in the tropics. He died at the family home surrounded by family members, Elizabeth and remaining children, Elizabeth Robson, Josiah Jnr, Charles and Annie. Well attended by townsfolk and locals in the district, a long cortege filed along Chilcott Street to the church where a Cornish-style funeral was conducted by Reverend C Dunstan, of Orange. Josiah was interred on his property in the lovely garden of the brick home he built along Oakey Creek. His wife Elizabeth died many years later on 26 March 1898 and was buried next to him in the garden, a fine marble tombstone marking the site.

Gravestone of Josiah Holman – repositioned in 2000.

Elizabeth Simmons (Simmy) Robson died 2 February 1895 and was buried at Cadia.

Josiah Jnr (Joe) died 23 December 1915 in Sydney and was buried in Waverly Cemetery.

Charles William stayed on at Rosemont until the properties were sold in 1906 and he and Ada and 11 children moved to North Sydney. Charles died on 19 April 1940 while visiting Orange and was buried in Orange Cemetery.

Annie lived at "Bexley", Four Mile Creek on the 460 acre property bequeathed by her father and died on 10 December 1953 aged 94.

Cadia Engine House restored – circa 2010

As for Cadia, it was but another chapter in the "book of gold fever" that engulfed the nation. It would be another century before Cadia rose once more to be added, not as a mere addendum to the book, but as a new chapter, one of the richest mineral deposits in the world, just as Josiah believed. Begun by Newcrest Mining Corporation, now Newmont, the company employs over 2000 people.

In his final report of June 1868, Josiah, applying his considerable hard-won world experience, wrote to the board of Cadiangullong Copper Mining Company, a report that may well be one of the most remarkable predictive observations of all time:

I have great confidence that the vein, abounding in iron and iron pyrites, traversing the whole length of the property, will ultimately be found to produce gold in payable quantity, and inexhaustible in its supply of auriferous stone ... some very good specimens of auriferous gossans (gold bearing rock) were found. Gold could be seen sparingly in this stone. This result is nearly, if not quite, payable by working it on a large scale. This reef warrants a deeper trial ...

That *inexhaustible supply of auriferous stone . .* is estimated to be one of the largest gold and copper deposits in the world.

Josiah's Last Will and Testament

THIS IS THE LAST WILL AND TESTAMENT of me Josiah Holman Grazier of Cadia, New South Wales, Australia.

After payment of all my just debts, funeral and testamentary expenses I bequeath as below:-

At my decease I bequeath to my daughter Annie Parish to be held in trust by my executors and after her decease to her children, Keen's Selections and additions thereto in all 460 acres, being the southern portion of Boxland Park, all fronting Cadia Creek with sufficient sheep to stock the land bequeathed. All my other real and personal property I leave to my executors to manage and to be held for the benefit of my wife during her lifetime only; and for the education and provision of my granddaughter Susan Louisa Blood, also for my eldest daughter Elizabeth Robson, and any other dependents of my family deemed by my executors to need help or provision. Of the above undescribed lands applied to the use of my wife during her lifetime; I bequeath that the decease of my said wife, to Susan Louisa Blood, my granddaughter, Boxland Park home paddock containing 107 acres and also the land known as Clark's and Flannery's Selections containing 80 acres, also 100 acres of land in addition adjoining the land on the east, also 40 acres formerly William Jenkins and adjoining on the North the beforementioned 100 acres, total amount of land 327 acres. I also bequeath to the said Susan Louisa Blood sufficient sheep to stock the land bequeathed. The above bequeaths to be held in trust by my executors for her use and benefit until she reached 30 years of age and thereafter to be held and administered by herself for her sole use and benefit.

I desire also that the said Susan Louisa Blood shall after the death of my wife reside with her aunt Annie Parish. I also bequeath after the decease of my wife, to my granddaughter Marianne Emily Blood the Ram paddock, North West of the Iron Duke, containing 97 acres, also Maroney's and Cummin's lots on Cadia Creek containing 80 acres, in all 177 acres. I also bequeath after the decease of my wife to my grandsons Frederick and Earnest Blood and to their heirs, the Washpool Paddock on Cadia Creek containing 260 acres, also selection 78/31 containing 103 acres, that is the eastern lot of Gum Flat, in all 363 acres. I also bequeath, after the decease of my wife, to my eldest daughter Elizabeth Robson, Tom Osborne's former selection containing 200 acres, in all 278 acres to be held in trust by my executors for her use during her life only. I also bequeathed after the decease of my wife to my daughter Annie Parish or to her children Lot 7889 containing 260 acres being additional selection adjoining the Home Paddock on the East, Boxland Park. I also bequeath after the decease of my wife to my son Josiah Holman or his children lot 76/85 containing 260 acres being Gun Flat proper. I also bequeathed after the decease of my wife Lot 84/96 containing 205 acres situated North of Gum Flat jointly to my daughter Annie Parish or her children; to my son Josiah Holman or his children; and my granddaughter Susan Louisa Blood or her children to be used as a stock run and for timber for fencing and fuel.

The residue of my property after my wife's decease and the 278 acres bequeathed to my daughter Elizabeth Robson, after her decease, I bequeath equally among my six sons and daughters or their children, excepting my piano, which I bequeath to my granddaughter Susan Louisa Blood or her children and if deceased to my daughter Annie Parish or her children. My Life Assurance Policy No 12314a with bonuses will suffice for immediate requirements for my wife, my daughter Elizabeth Robson and my granddaughter Susan Louisa Blood to be used by my executors as needed. I appoint executors to this my last Will and Testament my son Josiah and Charles Holman and my son-in-law Freddrick Parish …

Josiah Holman *born in Gwennap, County of Cornwall in 1821, baptised 14 October 1821. Died in Cadia 18 February 1893.*

Terminology

Adit – Horizontal or near horizontal tunnel or shaft used to extract ore.

Alluvial deposit – Sedimentary material, such as sand, gravel, and clay, which has been naturally deposited by running water.

Blast furnace – A large structure in which ore is smelted into metal.

Blister copper – A rudimentary form of copper (assaying about 99%) produced in a smelter requiring further refining before being used for industrial purposes.

Crosscut – A horizontal opening driven from a shaft and (or near) right angles to a mineral vein or other orebody.

Drift – A horizontal underground opening that follows along the length of a vein or rock formation as opposed to a crosscut that crosses the rock formation.

Elvan – Cornish mining term for dyke rocks of granite type composition containing quartz, orthoclase and tourmaline.

Excavation – Miners, using hand tools such as picks, shovels and hammers, would dig out the shaft while shoring up the walls with timber supports to prevent collapse.

Ferrous – Containing iron.

Flux – A chemical substance that reacts with gangue (worthless minerals) to form slags, which are liquid at furnace temperature and low enough in density to float on the molten bath of metal or matte.

Genesis – The origin or formation of natural minerals, its evolution taking millions of years.

Gossan – Rust-coloured capping or staining of a mineral deposit, generally formed by the oxidation or alteration of iron sulphides.

Granite – coarse-grained intrusive igneous rock consisting of quartz, feldspar and mica.

Killas – Cornish mining term for the altered alluvial rocks or clay-slates and siltstones which are generally found around granitic rocks.

Kibble – an iron bucket used for raising ore.

Limestone spar – a crystalline calcite that forms in spaces between grains in carbonate rocks such as limestone.

Lode – A deposit of valuable minerals contained within solid rock.

Nugget – A small mass of precious metal found free in nature.

Open pit – A surface mining technique where the mine is entirely on the surface, also known as open-cut or open-cast mining.

Ore – Naturally occurring material from which valuable minerals or metals can be profitably extracted.

Outcrop – A visible occurrence mineral deposit that is exposed on the surface.

Panning – Separating gold from other materials using a pan and water.

Pyrites – fool's gold, a naturally occurring mineral of iron sulphide.

Raise or Winze – a vertical or near vertical opening connects different levels of the mine mostly through the orebody.

Quartz – A mineral composed of silicon dioxide (SiO2), a principal component of igneous rocks and sedimentary rocks like sandstone and shale, and is also the principal mineral in sand.

Race or Tail Race – waterway or channel to move water on a decline

Serpentine Rock – fine-grained, compact rock that can be many colours.

Slag – The vitreous mass separated from the fused metals in the smelting process.

Sluice box – An elongated wooden or metal trough with ripples over which alluvial gravel is washed to recover minerals.

Stope – An excavation in a mine from which ore is, or has been, extracted.

Strata – A series of beds or seams of rock.

Tailings – The material or waste left over after the valuable product/commodity has been extracted from the ore.

Tied –bound in a common tie.

Tribute – The miners work on contract and tribute (payment), the price being fixed according to the quality of the ore raised.

Tut work – Miners engaged on wages (ie., 30s per week) to dig shafts and excavate, development drives.

Tunnel – A tunnel is an underground passageway, while an adit is a horizontal tunnel, and a shaft is a vertical entry passage used for access or to haul goods and a drive is horizontal or nearly horizontal passageway that follows a vein of ore.

Wolfram – dark grey to brown-black mineral found in granite rocks and quartz veins.

Acknowledgements

To those many family members and those across the ever-shrinking globe from country NSW, Cornwall, South Africa and Quebec who have assisted over the past years with this research I say thank you. I have contacted many local historians and historical societies in areas where he once spent some time and am grateful for the information that they were able to provide. They not only shared information about the notable people Captain Holman encountered, but also provided me with a glimpse into the lives and experiences of the early settlers during that time.

I am indebted to Fran Keable and husband for their tireless work in transcribing Captain Holman's expedition journal of Lower Canada, Margaret Tie for her continuing support and family information.

Also to my dear wife, an avid reader who assisted with additional research direction and kept me on track during my writing.

Source of Illustrations

I am grateful to all those who have supplied photographs and illustrations for use in this publication. I apologise if I have inadvertently infringed any copyright.

www.ingramcontent.com/pod-product-compliance
Lightning Source LLC
Chambersburg PA
CBHW061110070526
44583CB00027B/3245